The Pet Detective Series

IS A CAT THE RIGHT PET FOR YOU?

Can *You* Find out the Facts?

First published 2016

Published by

5M Publishing Ltd,
Benchmark House, 8 Smithy Wood Drive, Sheffield, S35 1QN, UK

Tel: +44 (0) 1234 81 81 80
www.5mpublishing.com

A Catalogue record for this book is available from the British Library

ISBN 978-1-910455-17-3

Book layout by Mark Paterson
Printed by Bell & Bain, Scotland

Photos and illustrations by Emma Milne unless otherwise credited

The Pet Detective Series

IS A CAT THE RIGHT PET FOR YOU?

Can *You* Find out the Facts?

By

Emma Milne

BVSc MRCVS

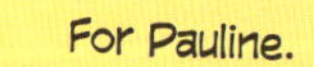

I'll think of you when I look at the stars because...

I like space too.

Contents

Acknowledgements

This book would not have been possible without the fantastic support, photos and help from lots of people. Thank you to the wonderful Animal Welfare Foundation and International Cat Care. Having endorsement from two such respected charities is a great privilege. International Cat Care, the use of your beautiful calendar photographs has really helped bring the book to life. Massive thanks to Claire Bessant, Andy Sparkes and Vicky Halls. Your help, advice, time and photos are deeply appreciated. Many thanks also to Dr Sarah Caney of Vet Professionals (www.vetprofessionals.com), Hill's Pet Nutrition, Eleanor and Scott Coultish, SureFlap Limited and Bayer Animal Health for the use of their great photos. Thank you to Emily Coultish for bringing my cartoon ideas to life and for putting up with my pedantry about details! I salute you all!

Chapter 1

THE JOYS OF LIVING WITH ANIMALS.

Humans and animals have shared this lovely little planet of ours for thousands of years. Over that time we've used and needed animals in all sorts of different ways. First of all we needed their bodies, and we used all the bits we could. We ate their meat and some of the other bits too, like kidneys and livers and hearts. We got rich nutrients out of their bones in the bit called the marrow. We used their stomachs to carry water in or to cook other bits in. Animal fat was used to make oils and candles and fuel lamps and torches. Bones and horns could be made into tools and cups and we used their skins and fur to keep us warm and dry. Even things like a bison's tail could be made into anything from a fly swat to an ornament.

These days we still eat plenty of animals, but we've found substitutes for some of the other bits. Over those thousands of years we started to realise that animals could be useful for other things. Cats were very good at catching mice and rats and other pests that ate our crops, so having cats around started to look like a good thing. They could do jobs for us that we found hard. Wild dogs started creeping closer to human camps to get some warmth from our fires and steal scraps of our food. In return, humans got some protection from their natural guarding instincts and we learned that if we joined dogs in hunting, we all made a pretty good team. Horses, donkeys and camels could be tamed and could carry big loads and cover distances that man couldn't even attempt. These animals are still used all over the world today.

As more time has gone by we've stopped needing animals so much to do things for us. We have cars and tractors, we have mouse traps and food containers that can't be chewed through, we have houses with alarms and strong doors and big locks and we have farmed animals to eat so we no longer need to hunt. But the simple truth is that we found out that animals are wonderful creatures, and in the time we were getting to know each other, humans started to fall in love with just having animals around.

Chapter 1

THE JOYS OF LIVING WITH ANIMALS.

A dog putting its head on your lap to have its ears stroked gets some lovely affection, but the human who's feeling those velvety ears and looking into those lovely brown eyes gets a lot back too. The cat stretched out on a warm rug by the fire has definitely 'landed on its feet', but the doting owner smiling in the doorway just watching the cat's tummy rise and fall with its breathing feels calm and happy without even realising it. Pets make us happy. They make us feel calm and loved and wanted. Pets don't judge us or hold a grudge for days like your best friend did when you spoke to the new girl at school. They are always there and they stick with you through thick and thin. In fact, sometimes they seem a lot nicer to be around than some humans!

The Serious Bit

The last point I want to make is the bit that is so important to remember: humans don't always do the right thing for their pets. Pets don't get to decide who buys them or how they are cared for. They have to live wherever you put them and they can only have the food you give them because they can't get to food themselves. When they go to sleep, how comfortable their bed is will be totally up to you. What you have to realise is that if you want a pet, those animals are *completely* dependent on you and your family to keep them happy, healthy and safe.

It sounds like an easy thing, doesn't it? Buy a cage, or a cat bed or a dog's squeaky toy, go to the pet shop and buy a bag of food and your pet will have everything it needs. WRONG! Thousands if not millions of pets have been kept this way, and of course with food and water most animals can survive for years, but that is just not enough. A great life isn't about coping or managing or *surviving*. It's about being HAPPY! If you were locked in your bedroom with no toys or books or friends or even your pesky sister, you'd *manage*. As long as your mum gave you bread and stew twice a day and the odd bit of fruit you'd probably live for years. But would you be happy? It doesn't sound likely, does it? You'd be bored out of your mind, lonely, miserable and longing for someone to play with, even if it was just that nose-picking sister or brother that you usually avoid like a fresh dog poo and make cry in front of your friends.

Having a pet, any pet, is a serious business. It's a bit like getting married or having a tattoo: you definitely shouldn't rush into it! You need to think carefully about lots of things. What sort of house or flat do you live in? How big is your garden if you have one? What other animals, if any, have you already got? How much money do your parents earn, because I can tell you there is no such thing as a cheap pet! How much spare time do you *actually* have? Are you an active family or a lazy one? All these questions have to be asked and they have to be answered very *honestly*. And of course you need to ask yourself what sort of animal do you want?

I've been a bit sneaky there, because actually you should *never* ask yourself what sort of animal you want. You should think about what sort of animal you can look after properly. There's a very famous song from a long time ago called *You Can't Always Get What You Want*, and I'm sure your mum or dad will have said it to you hundreds of times. You probably rolled your eyes, walked off in a huff, slammed a door and shouted, 'That's not fair!!' But I hate to say that your mum and dad are right, and it's especially true when it comes to keeping pets. Lots of animals get abandoned or given away because people don't ask themselves the right questions, don't find out the facts and then, most importantly, don't answer the questions truthfully.

Let's be honest, you lot are masters at pestering. For as long as children, parents and pets have been around, children have pestered, parents have caved in and pets have been bought on an impulse! This means without thinking and without knowing what the animal actually needs to be happy, which usually means a very miserable pet. But we're about to change all that, aren't we? Because now I've got the dream team on my side. You chose to find out the facts about these animals so you *could* make the right choice. And I am very proud of you for that and I am very happy. So thank you.

The *EVEN MORE* serious bit!

So, you are thinking it would be nice to have a pet. You're certain you are going to love it, care for it, keep it happy and, of course, *never* get bored with looking after it and expect your mum and dad to do it. But what you need to know is that not only is that the right thing to do but it is also now the law. Sounds serious, doesn't it, but as I said, it's a serious business. In the United Kingdom in 2006 a new law was made called the Animal Welfare Act. This law says that anyone over the age of 16 looking after an animal has a 'duty of care' to provide for all the needs of the animal. Now, laws are always written by people who use ridiculously long words and sentences that noone else really understands, but this law is very important to understand. A duty of care means it is the owner's responsibility to care for the animal properly and the law means that if the owner doesn't, they could get their pet taken away and even, in rare cases, end up going to prison!

Aha, you may think, I am not 16 so I'm fine, and you'd be right, but the duty of care then falls to your mum

Chapter 1

THE JOYS OF LIVING WITH ANIMALS.

or dad or whoever looks after you and the pet. So if you would like a cat, not only do you need to know all about them, but you need to make sure the adults in the house do too. And you need to make sure they know about the law, because they might not know what they are letting themselves in for!

If you don't live in the UK you need to find out what laws there are in your country about looking after animals. But remember, even if your country doesn't have any laws like this, making sure your animals are healthy *and* happy is still simply the right thing to do.

Well, that's quite enough of all the boring serious stuff – let's learn some things about animals! The easiest way to find out about animals is to know about the five welfare needs. These apply to all pets, and in fact all animals, so they are good things to squeeze into that brilliant brain of yours so you can always remember them whenever you think about animals.

The Need for Fresh Water and the Right Food

This is a very obvious thing to say but you'd be surprised how many animals get given the wrong food. In fact, there was once a queen a very long time ago who wanted a zebra. I said wanted, didn't I? She definitely didn't ask herself the right questions or find out the facts, because when someone caught her one from the wild, she fed it steaks and tobacco!

Animals have evolved over a very long time to eat certain things and if they are fed the wrong foods they can get very ill, very fat, or miss vitamins and minerals they might need more than other animals. The right food in the right amounts is essential.

The Need to Be With or Without Other Animals

Some animals live in groups and love to have company. Some animals are not very sociable at all, like me in the mornings! It's very important to know which your pet prefers. If you get it wrong you could have serious fighting and injuries or just a very lonely and miserable pet.

The Need for the Right Environment

This is a fancy way of saying where the animal lives. It could be a hutch, a cage, a house, bedding, shelter, a stable or lots of other things, depending on the pet. It's very important that their homes are big enough, are clean, safe and secure and that the animals have freedom to move around.

The Need To Behave Naturally

Knowing what animals like to do is really important. As we said before, lots of animals will survive on food and water, but happiness or 'mental wellbeing' is just as important as being healthy or having 'physical wellbeing'. You've probably never thought about your own behavioural needs, but imagine how you would feel if you were never allowed to go to the park or play or run or see your friends. You would soon be quite unhappy. Often you find that happy pets stay healthier, just like us.

The Need to Be Protected from Pain, Injury and Disease

Animals can get ill just like us and it will be up to you and your family to keep your pet healthy as well as happy. Just like you have vaccinations, they are very important for some animals to stop them getting ill and even dying.

Animals, just like lots of children, also get worms, lice, mites and other parasites. You will need to find out how to treat or prevent these and look out for signs of them.

You need to check your animals over at least once a day to make sure there are no signs of problems and take them to a vet as soon as you think something is wrong. Vet costs are not cheap. You might also have a pet you can get health insurance for, which is always a good idea.

So now you know the basic needs of all animals, it's time to concentrate on cats! Cats may have lurked around us humans for a long time, but they still have lots of wild instincts and needs. The best way to learn what will keep your pet cat happy and healthy is to find out what cats in the wild are like. How do they live? What do they eat? Do they like to have others around and what makes them scared or nervous? In other words, what keeps them *happy*? Shall we begin?

Hannah Winser c/o International Cat Care.

Chapter 2

CATS IN THE WILD.

From big cats to kittens, cats have more in common than just being beautiful, sleek creatures. They're great climbers, they have big sharp teeth and very sensitive ears and whiskers. But there are also lots of ways they've evolved to be different. For example, lions are virtually the only cats that hunt in a group and tigers are the only ones that seem to like water!

Looking at how our pets started out and how they live *without* human beings getting in the way is the best way to understand how to keep them happy when we do interfere!

Pet cats are descended from desert cats like the African wild cat and started to hang out with humans a very long time ago. Maybe even as long as 10,000 years ago! As we mentioned before, cats are pretty useful to have around if you've got crops and rodents like rats and mice in the same place. Rats, mice and voles can be a real pest for humans and eat mountains of our food. When humans started to store food, we soon became very attractive to these little beasties. As the crop stores got bigger and the numbers of rodents grew, our villages started to look pretty good to wild cats as an easy place to get a juicy meal. Humans soon realised that it was handy to have the cats around to keep the pests under control and this is how cats and humans started to get along.

Shutterstock

Lions live and hunt in groups.

Shutterstock

Tigers are the only cats that love a dip!

Shutterstock

A beautiful African wild cat.

CAT FACT:

The ancient Egyptians thought cats were so fabulous they believed they were gods. Some Egyptians even had their pet cats mummified so they could stay together for eternity. On the other hand, later on in Europe when everyone was worried about witches, one of the popes decided that all cats were "diabolical". This means he thought they were friends with the devil. Sadly, for the moggies this meant that lots of them got killed for absolutely no reason!

Luckily we came to our senses and realised that witches didn't exist and that cats are actually brilliant animals to have around, and our love of cats has grown ever since. Over the years, pet cats haven't really changed at all because they were already great at what we needed, and loved killing things, as well as being fascinating to watch and really lovely to look at! This means that even though cats have been around humans and even kept as pets for thousands of years, they still have lots of the same instincts and needs as wild cats.

One of the best ways to understand animals and virtually everything about them is to find out about what they like to eat and what, if anything, eats them. Every animal on the planet has to eat to survive, so food and water are the most important things to them. Animals in the wild never really know when their next meal might be, and getting food can be really hard for them. Let's face it, they can't exactly order a takeaway when they're tired after a hard day!

Chapter 2

CATS IN THE WILD.

Over the millions of years that life has existed on our beautiful planet, animals and plants have evolved together in an amazing balance of nature, all based on food chains. Food chains are the way we look at which animals eat what. For example, a simple food chain for a cat might be:

In this food chain the mouse is what we call the prey because it is eaten by the cat. The cat is the predator because it eats the mouse. Predators and prey have usually evolved very differently because of the ways they need to get food. If you're a mouse, food might be easy to find, but you always have to be on the lookout for what might be about to eat you! The mouse, like lots of prey animals, has eyes on the side of his head so that he can see all around and above him all at the same time. He needs to be very alert and also very quick to escape if a predator comes along.

On the other hand the cat needs to be cunning and quiet. He needs to outsmart the mouse and creep up on him without being heard or seen. He also needs to have great senses to find the mouse in the first place.
Most predators have eyes right on the front of their face, like humans, cats and dogs. This means they can't see behind them like mice but they can judge distance and depth really well for catching fast-moving things. Don't forget though that even though teachers are humans and have forward-facing eyes, they can still see what you are doing behind them without moving their heads!

Of course, when you think about it, lots of animals are predators and prey at the same time. Humans are one of the most successful predators on the planet but there are still plenty of things that can make a meal of us given the chance:

Cats are just the same. Big cats like lions might not have much to worry about but smaller cats need to be on the lookout just as much as any prey animal. Cats can get chased, injured and killed by lots of animals like dogs, foxes and even other cats. So you start to see that there are certainly lots of times that cats might feel frightened too.

Different cats live very differently. We said that lions hunt in groups, but most wild cats definitely don't. Cats in the wild also never share food, apart from when a mum shares with her kittens. They need to catch lots of little animals and birds every day just to survive. They don't really have particular times of day or night when they are most active and need to catch 10–20 small animals a day to survive.

Prey animals are very good at escaping and use all sorts of ways to avoid being eaten, so life can be very hard for predators. This means that even if a cat is halfway through a delicious rat and a cheeky mouse ran past it, the cat would still try and catch the mouse without even thinking about it. They can never afford to miss an opportunity to get food.

CAT FACT:

Cats have retractable claws. This means they can choose when they are tucked in and when they are out. This is really useful because they can hide them away most of the time so they stay nice and sharp and so they don't make a noise like a woman in high heels when they are trying to sneak up on prey!

Shutterstock

Claws can be hidden away...

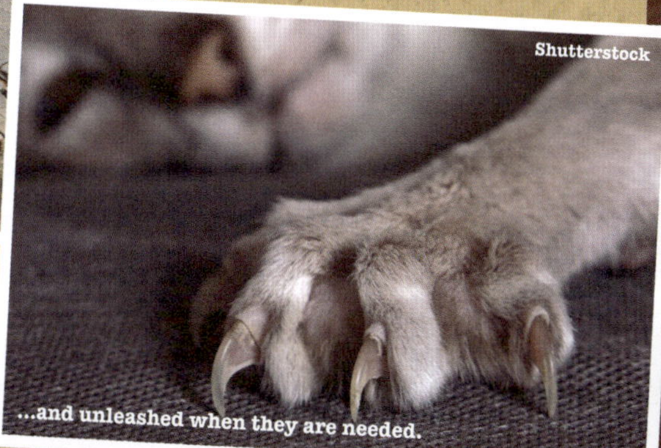
Shutterstock

...and unleashed when they are needed.

Chapter 2

CATS IN THE WILD.

Because of the way they live and hunt and because they are about as good at sharing as a selfish toddler, cats in the wild like to avoid each other. They have a territory that is their own and they get very upset if another cat tries to get onto it, because they want and need all the food to themselves. They spray urine (wee) round their territory in obvious places and they scratch trees and fences to leave their scent from their paws. This is their way of hanging a "KEEP OUT!" sign on their land. Some cats like a huge territory of acres and acres to keep them from going hungry, so battles between neighbouring cats can easily happen.

Sensitive Creatures

Cats really are very sensitive creatures. I don't mean they will easily cry at a girly film but they do have amazing senses. It's important to know how sensitive they are, because our world can easily be totally overpowering for them.

Cats have eyes that face forwards and can see really well in low light.

Shutterstock

Vision

Cats can't see in the pitch black, BUT cats have much better vision than us in low light. Lots of prey animals try to avoid being eaten by coming out when it's as dark as possible. This means cats could have more chance of catching something if they laze around all day and come out at dusk and dawn.

Cats' eyes are very good at using every scrap of light and they have a layer at the back like a mirror to help reflect as much light as possible. The way a cat's eyes shine at night is what gave the man the idea for the reflective bumps in the road. And of course, why the bumps are called cats' eyes!

Cats' eyes might be great in dim light but they are not very good at seeing colour and they also can't focus on things very close to them like humans can. They rely on being very good at spotting quick movements and pouncing.

Smell

We said that cats use urine and scents to mark their territory, so smell must be really important to them. They can smell when they are in a safe place by recognising their own smell but also will realise when they've strayed into someone else's place. When cats do live with other cats, they will use their sense of smell to help recognise the other cats and also the people they live with.

Cats have noses almost as powerful as a dog's. You can imagine that some of the smells we have in our lives could seem like an unbearable stench to such a sensitive creature. Think how you feel when you get grabbed by your aunty who's drenched in a fug of bad perfume, and pulled in for that awkward hug and a sloppy kiss. An unpleasant experience for you could seem like an act of war to a cat!

Hearing and Touch

Cats can hear a really wide and unusual range of sounds, from the high-pitched squeaks of their prey to the low rumblings of human voices. This is rare in most animals, so busy houses full of kids, dogs and music might seem pretty deafening. Their ears have help from *20* muscles to help move them to just the right place to hear where squeaks are coming from.

Shutterstock

Cats have mobile ears and sensitive whiskers.

Cats also use touch a great deal. Their feet and whiskers are very sensitive to the feel of things and to vibrations that could mean a tasty mouse is hiding somewhere.

Taste

When it comes to taste buds, cats are *not* like most kids. Sugar and salt do nothing for them. In general, animals tend to be one of three things. Animals like cows, horses and rabbits are herbivores and just eat plants. Animals like humans, bears and dogs are omnivores and eat animals *and* plants, and animals like cats, birds of prey and mink are carnivores and virtually just eat animals. Animals' bodies don't taste sweet or salty, just meaty, and that's the taste cats like best.

In the wild, cats need to eat the whole animal (apart from the stinky gut contents) to get a good balance of all the nutrients they need. We'll look at this more in the next chapter but for now it tells us that cats have no interest in sweet foods, and if you want a vegetarian pet, then cats are definitely not for you!

Chapter 2

CATS IN THE WILD.

CAT FACT:

Cats have hundreds of tiny, stiff barbs on their tongue. If they lick you, you will feel how rough the tongue is. These help rasp all the last scraps of meat from the bones of their prey. They also help cats stay super clean. Cats love to keep their fur very clean and spend lots of time grooming. Their rough tongues are like having a loofah with them all the time!

Shutterstock

Shutterstock

All cats have barbs on their tongues but some are scarier than others!

Life In 3D

Animals see the world in a very different way to us, and cats are no exception. Cats are predators but still need to escape from things, and they are amazing climbers and jumpers. They see every nook and cranny of their surroundings. We walk into a room and we tend to focus on what is at eye level or lower; cats will see everything else. They like to hide when they feel threatened and they like to tuck themselves away when they need peace and quiet. Kittens will stay low and hide away, but adult cats will always get up high whenever they can.

Shutterstock

All cats love to climb

Shutterstock

So now that you know pretty much everything there is to know about cats in the wild and where they came from, we'd better get down to the tricky business of keeping them as happy, healthy pets. In the meantime you might start to think about how your house looks, smells and sounds from a cat's point of view.

Chapter 3

FRESH WATER AND THE RIGHT FOOD.

My cheeky cat Brian, when he was a mischievous kitten.

We've already said that a great life is what we should try to give all our pets, and the best start to that is to get the basic survival stuff exactly right from the very start. The top three things needed for life are air, water and food. Animals, including humans, can't live without these things, and when it comes to food, getting the right diet and feeding the right *amounts* of food will get your cat off to a brilliant start for a healthy and happy life.

Let's tackle the easy part first. Water. Water is absolutely essential for every living thing on Earth. For animals like humans and cats, after the need for air, water is the most important thing. If animals can't get to enough water, they can get very ill and die really quickly. Water is the only thing your cat needs to have to drink. Depending on the food you give them, they will also get some water in their food, say for instance in pouches or tinned food, but it's essential they have access to plenty of fresh water all the time as well. Just like us, they will need to drink more during warm weather compared to cold weather and also if they've been very active, running, exploring and playing.

The easiest way to give your cat water is in a bowl. The bowl should be easy to clean and be wide enough that your cat can have a drink without touching its sensitive whiskers on the edges. Cats also like bowls to be very full so that they don't have to put their head right down to have a drink. This way they can keep a careful lookout for dangers, just like their wild relatives.

Some cats love to drink running water, like a dripping or running tap or out of streams. Sometimes this is because they like running water, but some cats also don't like the smell of tapwater because we treat it with chemicals to make it clean for humans. You might also find your cat is cheeky and drinks from your glasses as well!

Chapter 3

FRESH WATER AND THE RIGHT FOOD.

You can also get all sorts of fountains these days to offer your cat different ways to drink, but usually a bowl is fine. You'll need to make sure the water is clean and change it twice a day so it's always fresh. Our cat, Brian, always used to paddle in his water bowl before he had a drink. This meant his water got dirtier than most, so if you end up with a real character like Brian you'll need to watch out even more!

iStock

Lots of cats love drinking from a running tap...

iStock

...and some will cheekily steal from your glass!

So, first we need air, then water and then of course we need food. Bodies are amazingly clever, and make any computer or machine that humans have built look like the most rubbish toy imaginable. Like all machines, bodies need energy to make them work. Because our bodies are so spectacularly clever, they can also fix themselves and replace their own parts. Instead of plastic and metal bits or replacement batteries they need vitamins and minerals, fibre, proteins, fats and carbohydrates. All animals get these from their food, and over the millions of years that animals have been on the planet they have evolved different ways to get all the things they need from their food. As we said, wild cats eat lots of small animals all through the day and night.

Carnivores such as cats that eat practically nothing except other animals are very good at making energy from proteins.

CAT FACT:

Cats aren't just carnivores, they are 'obligate' carnivores. This means that in the wild they *have* to eat meat to get all the essential things they need. If cats don't get these nutrients from their diet, they can very quickly die.

So what do we feed pet cats? Are you going to have to go into your garden 15 times a day and catch mice and birds and then feel really sad when you feed a *live* animal to your cat? NO! Of course not! Phew.

The great thing is that we live in a time where we know absolutely *loads* about what animals need to eat and what they need to grow perfectly, live healthily and then grow old well too. All the work has been done for you and there are dozens of brands of cat food you can choose from. Now the only thing you have to do is pick which one to feed! If you do get a cat or kitten, have a chat with your vet right at the start or, even better, *before* you get your cat to find out what food they recommend and why. Most cat foods will provide everything your cat needs, but some are better than others. Most vets recommend what we call 'super premium' foods. This means they don't just meet your animal's basic needs, but they have good quality ingredients, they are very strictly controlled so they are perfectly balanced and they have the little extras in them for tiptop health.

Growth

Growing up is hard work and it takes lots more food and nutrients than being an adult. You may have heard your mum or dad say they wish they could eat as much as they used to. When I was a little girl I used to eat about eight Weetabix a day as well as all my normal food. My bones and muscles were growing, my brain was developing and I was always running, skipping, jumping, doing cartwheels and climbing trees. Nowadays I eat about half what I used to because I'm not growing anymore, I don't have the energy to run about and last time I tried a cartwheel I pulled a muscle!

Growing animals (and kids) need more calories or energy, more protein and more of certain minerals than adults. They also need things called fatty acids to help their eyes and brains develop as best as they can.

Elahe Soufani c/o International Cat Care.

Kittens need lots of energy and nutrients for playing and for growing brains, muscles and strong bones.

Chapter 3

FRESH WATER AND THE RIGHT FOOD.

It's really important to feed kitten food to kittens. It sounds obvious, doesn't it, but lots of people don't. You might find your kitten would survive on adult food, but if you want it to grow properly and be the healthiest, brainiest cat it can be, you need to feed it good quality kitten food.

This is a good place to talk about dry food and wet food. Just as it sounds, dry food is the dry kibble biscuits you get and wet food is the meaty-looking food you get in pouches and tins. Some wet foods are chunks in gravy and some are more pâté-style tinned foods. Virtually all foods are 'complete' now, which means they provide everything your cat needs, but do check that whatever food you pick is complete. In general it doesn't matter if you feed wet or dry because both should provide everything.

BUT cats are creatures of habit and if they have dry food only or wet food only they can get hooked on the texture and might never eat anything else. This may never be a problem for you but there are times in a cat's life where you might need to change its food. Wet food tends to work out more expensive, so if you found your money situation changed, you might need to switch to dry food. Just the same, if you have a cat hooked on dry food and they end up with certain health problems, like kidney problems or bladder problems, your vet might prefer them to have wet food.

Like being a good Boy Scout, if you feed a combination of wet and dry food from the start, you can be prepared for times where you might need to favour one over the other. It's not essential but it's a very good idea if you can manage it. It's also a nice bit of variety for your cat. Of course if you do the great good deed of adopting a cat from a rehoming centre, you might not get the choice, but it's worth trying a combination and seeing if your cat can adapt. Some aren't as set in their ways as others!

Hill's Pet Nutrition.

Try to feed some wet food...

Hill's Pet Nutrition.

...and some dry food

So, logically, if we feed kitten food to kittens, we should feed adult food to adults and mature cat food to older cats. These various foods might be called different things and different brands might have different age guides, so just check with your vet about what to feed when.

Just as important as feeding the right food is feeding the right *amount* of food. Lots of animals have really different calorie needs. This is how much energy they need. Some cats are very active, especially when they're young, while some love to sleep 20 hours a day and barely move from the settee. If you feed both these cats the same amount of food, the very lazy one could quickly start to get way too fat.

Being too fat, or obesity, as it is called, is a real problem for animals and humans alike. You don't find fat animals in nature. Some animals will build up stores of fat to keep them warm and give them energy through the winter but you will never find a truly fat animal in the wild. Being too fat can give lots of animals diseases and the extra weight puts strains on joints and bones and gives the heart too much work to do. You can imagine that a fat predator would never manage to catch any prey, so would never get fat in the first place. And a fat prey animal might struggle to get away from a predator and wouldn't live long enough to have fat babies. This really is survival of the fittest!

Chapter 3

FRESH WATER AND THE RIGHT FOOD.

As well as the health problems, being overweight can also make it very hard to keep clean. Lots of animals need to groom themselves to stay clean and keep their fur and skin healthy. Being too fat makes this really hard, and for wonderfully clean animals like cats this is really horrible and very frustrating. We'll look more at these problems in Chapter Seven, but for now all we need to know is that the right diet and the right amount of food is really important.

Most cat foods have a guide on them to give you an idea of how much your cat will need a day. As different cats vary so much, these really are only a guide, so you need to watch out for if your cat is too thin or fat. It's very useful to know about something called body condition score, or BCS. This is a way people like vets judge how fat or thin or just right an animal is. We give a score out of 5, where 1 is dangerously thin, 2 is underweight, 3 is just right, 4 is overweight and 5 is dangerously obese. Have a chat with your vet or vet nurse to show you on your cat what to look and feel for. If you stop them getting fat in the first place, it's much easier than trying to get the weight back off again!

Our cat Belle learning about all the birds she might chase!

The last thing to tell you about feeding cats might surprise you until you think back to our wild cats. Most humans feed cats twice a day. Do you remember how many little meals they have in the wild? That's right, 10-20! Because of the way cats naturally eat, they have quite small, simple stomachs and like to have something to eat every few hours. Now you might not manage to give your cat quite so many meals as this, but if you can manage say six to eight meals a day, your cat will be very happy. You can buy timer feeders to help you if you're out at school or your mum and dad are at work. These have lots of compartments and open up when you set them to, so you can spread out your cat's food through the day, and even through the night if you want to.

Remember you still need to just split up their day's ration, not feed them six times as much! Don't worry if the meals look small. That's what they're used to!

Now we know all about what pet cats need to eat and drink let's look at what they need next. You'll see as we go along that the next three needs are all closely linked: the need to be with or without other animals, the right environment and the need to express normal behaviour. Shall we get started?

Chapter 4

THE NEED TO BE WITH OR WITHOUT OTHER ANIMALS.

Adrian Tan c/o International Cat Care.

This is probably the need that humans have messed up the most over all the time we've kept animals as pets. For hundreds of years we've kept rabbits and guinea pigs on their own, rabbits *with* guinea pigs, left dogs alone for hours while we're out and got used to the 'crazy cat ladies' who have dozens of cats in a tiny house. As you may have guessed, ALL THIS IS WRONG!

If you're shocked about rabbits, guinea pigs and dogs, then you'd better get those Pet Detectives books next, but for now we must focus those brilliant brains of yours on cats.

As we said so many times before, the best way to make sure our pets are happy is to look at their wild relatives and how they evolved and liked to live. Left to their own devices, cats are solitary creatures. They hunt alone, they spend most of their time alone, they hate sharing anything and they can easily feel scared by things because they are quite small.

Over those 10,000 years that cats started to hang around with us, they've gradually got to know humans and even like some of us! We keep them warm, dry and well fed and we have comfy beds and settees. BUT they still see other cats as possible threats to their territory and they are still frightened of lots of other animals, people and things.

Chapter 4

THE NEED TO BE WITH OR WITHOUT OTHER ANIMALS.

Hundreds and thousands, maybe even millions of people have more than one cat in their house. Lots of these people have neighbours who have cats. We live close together most of the time and cats love to wander, explore and patrol their territory. This means that lots of pet cats are forced to share space with other cats, both at home and in the neighbourhood.

Cats like to be the only king of the castle.

Cats are one of the most popular pets in the world now and people who love cats often think it would be lovely to get more and more cats because they love them so much. You'll also find that lots of people assume cats will get lonely if they don't have a cat friend. This is what's called anthropomorphism. That's a mouthful, isn't it!? It's when humans assume that animals are just like us and have the same needs as us. Doing this can be dangerous because, as you brilliant detectives are finding out, animals can be *really* different from us.

But we are going to change the world, aren't we? The way to do that is to stop doing that ridiculously long word and do two other things instead. Research and empathising. *Research* is what you're already doing by reading this book. Finding out what makes animals happy will help you be fantastic pet owners. And luckily for you, kids are usually amazing at empathising too, so you won't even have to try!

Empathising is imagining how you would feel in a similar situation to someone or something else. For example, if you saw an animal that had been starved, you could empathise because you know what it's like to feel hungry. If you saw a cat get chased by a dog, you might know how you would feel if the school bully came after you. Your heart would be pounding, you'd be frightened and you'd want to run away.

Humans are social creatures, like rabbits, dogs, horses and guinea pigs. They like to be with animals the same as them. That's how we, and they, feel safest and happiest. But cats are not like this, and for a social creature like a human, that can be hard to understand or empathise with. Cats are very happy with their own company and don't need other cats to feel happy. As we'll see in Chapter Seven, living with other cats they don't like can even cause lots of health problems for some cats.

There are times when cats have to live together these days. Modern day cats don't always end up with a loving home and there are thousands of cats living semi-wild all over the world. We call them feral or semi-feral. They're still the same as the moggie you might have as a pet, but they don't rely on or trust humans to survive. Cats are still good hunters and can often manage without us. But because there are so many of them living like this, food can sometimes be difficult to find and feral cats can often be found where some kind person is putting cat food out for them.

In these situations cats do sometimes live in a group and can even make friends with other cats. They are kind of forced into being friends or at least tolerating each other because they need the food.

The big difference between this kind of group and a house full of cats is the freedom and the choice they have. Any of the cats can come and go as they please. If they feel uncomfortable or stressed, they can leave the group and go off on their own. Lots of cats kept in a house together never have that option and that's when they can be really sad and miserable. You might be able to empathise with them if you've got a really annoying brother or sister but you're both too young to legally move out!

Chapter 4

THE NEED TO BE WITH OR WITHOUT OTHER ANIMALS.

You might know people with more than one cat and they might think that their cats are completely happy living together. The problem is that cats are very subtle creatures. This means they don't show emotions in big obvious ways like dogs and humans. If your brother is picking on you or pulling your hair or breaking your toys, you'll shout, 'Muuuuuummmmmm!' or storm off in a huff and slam your bedroom door. Which is, by the way, exactly what will give your brother that victory smile that annoys you even more!

Cats bully and get bullied in quite quiet ways most of the time. Cats being picked on won't scream and shout and slam doors. They will tuck themselves away, they might obsessively start grooming to try and calm themselves down and sometimes, you'll be amazed to learn, they will actually pretend to be asleep!

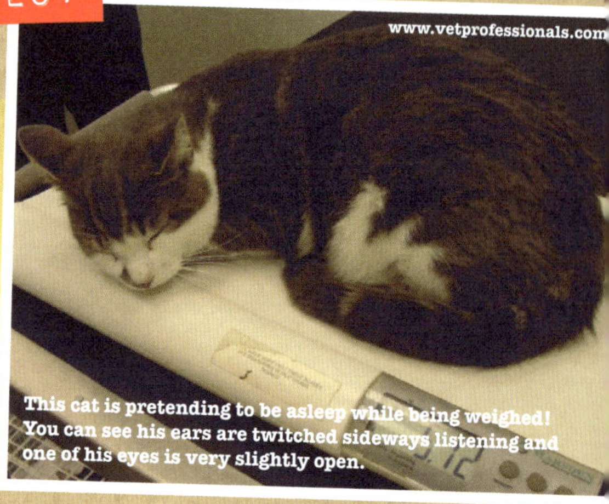

This cat is pretending to be asleep while being weighed! You can see his ears are twitched sideways listening and one of his eyes is very slightly open.

This cat is also pretending to be asleep. One of his ears is at an angle because he's listening hard to what's going on.

This cat, on the other hand, curled up, head half upside down and body totally relaxed, is away in dreamland fast asleep. Difficult to spot the difference, eh?! But now you know what to look for.

The cats doing the bullying won't always attack or bite or hiss, they might just sit near the food bowl or the litter tray and stare at the other cat. This might be hard for us to understand but it's important we know what to look out for so we don't keep misunderstanding our cats. At work when I talk to people about how their cats might be stressed or unhappy, they will often tell me they can't be unhappy because they never fight and one of them sleeps all day. Now you know what's *really* going on in that house!

Sometimes cats can live together and be firm friends. The way to know if two cats are true friends is that they will groom each other and often cuddle up together. The cats that are most likely to do this will be littermates or cats that grew up together from being kittens. So you could get lucky and have two happy cats, but there is never a guarantee. Sometimes you'll find that kittens will be really happy together, but once they grow up they might simply hate being near each other. It's a bit of a gamble, just like human brothers and sisters!

Anita George

Cats that are true friends will groom each other...

Louise Woodbridge

Sally Murfitt

...and snuggle up tight together

Chapter 4

THE NEED TO BE WITH OR WITHOUT OTHER ANIMALS.

Shutterstock

A simple tree is practically an entire playground...

The easiest way to keep things simple is to just have one cat. This way you can devote all your love and attention to one cat and your cat will have the best chance to be happy and have a calm and stress-free life. They will have the house, garden and maybe beyond to themselves, and if there are cats in the neighbourhood, they will still have the choice to avoid them if they want to by staying indoors. And most importantly, like that selfish toddler we mentioned, they won't have to share their food, bed, toys or litter tray with anyone else. What we're actually saying is that they will have their environment to themselves, just the way they like it. So now we'd better talk about what that environment should be like, shouldn't we?

Chapter 5

THE RIGHT ENVIRONMENT.

EASY! YOUR PLACE!

Hang on a minute: maybe we should investigate a bit more than that. Now we know what complicated creatures cats are, it's probably not as simple as we imagine.

Kittens are very good at getting into trouble!

Phil Croucher c/o International Cat Care.

One thing you need to know is that cats like things on their own terms. This means they think they are the boss and it means that it's very hard to get cats to do things you want them to do. They decide what they want to do and they tend do it whether you like it or not. If you wanted a pet you could train and order about, then a cat is not for you! In fact there's a very common saying that goes, 'Dogs have owners, cats have staff.' If you take a cat on, you need to be prepared to be *its* slave, not the other way round!

Remember what we said about cats living life in 3D and wanting to be high up? Well, this is important to know because it means that your cat will not mind where it sits and it won't care what you or your mum or dad think about that either. They love sofas, windowsills, beds, the top of the fridge, the airing cupboard, pillows, jumpers, the dining room table and usually your mum or dad's computer, work stuff and newspaper!

Chapter 5

THE RIGHT ENVIRONMENT.

Beautiful Woody used to visit us a lot and loved to sit right on my work stuff to make sure I noticed him!

Eleanor and Scott Coultish.

Clean washing always looks really comfy!

Eleanor and Scott Coultish.

Going on holiday? Make sure you don't pack anything unexpected!

If you share your home and life with a cat, you need to make sure your family are pretty chilled about a cat sitting on most of your belongings. They will also love to climb and explore bookshelves and mantelpieces, so precious ornaments might be in danger of getting broken. Kittens and 'teenager' cats can really tear about the place, climbing curtains, racing up and down stairs and generally getting into as much mischief as possible. This is way more entertaining than watching TV but might not put a smile on your mum's face!

The high places thing is not just something they want to do, but it is one of their needs and it is up to you to make sure they have such places. Some houses these days are very bare and uncluttered. This might be lovely for the humans living in them but it could be really boring or even frightening for a cat. Have a good look at your house from a cat's point of view. If you don't have lots of high places, you might need to think about ways to make some. Like putting a chest of drawers next to a wardrobe so your cat can leap from one to the other. A lovely soft jumper on top of the wardrobe that smells of you would make a great bed up there too.

Cats *love* high places

Of course you can buy all sorts of lovely beds, but you will often find cats like to pick their own sleeping places. There are plenty of people who spend a fortune on a plush cat bed, only to find it totally snubbed and unused. Life on their terms, remember! Whether you use a soft, old jumper or a proper cat bed, think about where they go and you'll have more chance of it being used. Adult cats like to be high and sometimes they like to be hidden. They love warm places. Put a few beds round your house in sunny spots on the landing, on cupboards or non-slippery shelves or on your own bed if your mum and dad say it's OK. Once your cat gets to know you, it will love the familiar smells of you, your family and the house to make it feel safe. With this in mind don't wash its beds *too* often, because cats like them to smell familiar.

Of course, even if you give them loads of beds, sometimes they still want to share!

Talking of beds, you do need to make sure your cat has places to sleep, so you'll need to think about beds carefully. Cats spend a huge amount of time resting and sleeping, so they need to feel safe, secure and confortable. They also *love* to be warm.

If you've got a kitten, remember that they can't climb well to start with so will like to hide under things. Make sure they have sleeping places tucked away, such as under a bed, so they can get peace and quiet and feel safe. Kittens and shy cats might like a bed with a roof so they can hide away in the dark and just peek out when they feel ready.

Kittens hide under things because they can't climb well.

Chapter 5

THE RIGHT ENVIRONMENT.

Just like all animals, your cat will need food and water bowls and somewhere to go to the toilet. With all these important things, you need to remember that cats don't like to share, so if you do have two cats, you need at least two of everything, preferably three, and they need to be away from each other. This way if your cats aren't as good friends as you think they are, they don't have to get too close to get all the things they value most. Even if you just have one cat, it's nice to have two or more of everything in separate places so that if they are feeling shy or you have visitors, they can still go somewhere quiet to eat, sleep or have a wee!

On the subject of going to the loo, what would you think if your mum decided to give you your dinner in the bathroom? Do you think it would be nice to eat a meal in there? Maybe it's nice and clean but maybe someone's done a big smelly poo in there and forgotten to flush it away. Maybe it smells of really strong cleaning chemicals, which is almost as bad sometimes.

The point is that loads of people get a cat and then put their food and water right next to their litter tray so everything is neat and tidy all together. Please don't do this. It's just like you having tea in the loo — not very pleasant.

Now imagine you finished your tea, needed the loo and you found that someone had taken your toilet and put it in your living room by a big window to your garden. Would you be happy to have a wee in full view of all the people and traffic going past your house? Probably not. So *empathise* and put your cat's litter tray somewhere nice and quiet and private away from the hustle and bustle of your busiest rooms.

Not all cats will need a litter tray. One of the great things about cats is how clean they are and the fact they can go outside to go the toilet on their own and you won't need to clear it up. Cats always dig a hole to have a wee or poo and then cover it up afterwards. Most cats that can get outside will choose to go to the toilet out there so they can pick exactly where and when to do it.

You'll need a litter tray (or more than one) when you first get a cat or kitten because they'll need to stay in until they're used to their new home or until they are old enough to go out. It's always useful to keep a litter tray in case your cat prefers it or at times when you might need to keep it indoors like when you move house or before a trip to the vet's. In general, though, most cats won't need cleaning up after on the toilet front.

Now you might notice I said there are times you *might* need to keep your cat indoors. More and more people these days, especially in busy cities and in places like the United States, actually plan to have cats that stay indoors all the time. In Europe most cats are still allowed to come and go as they please. So what's the right thing to do?

Well, the answer to this probably depends on lots of things: you, your family and where you live, but most importantly it depends on your cat and its individual personality. You might decide you don't want to risk anything happening to your cat, like being run over by a car, so you'd rather keep it inside. You might live somewhere in the world where there are laws about whether your cat is even allowed outdoors. You might end up with a cat that is really scared of everything in the outside world and that definitely doesn't want to go out even if it had the chance.

BUT when we think back to how cats have evolved and how their wild relatives live, you might start to feel a bit uncomfortable about keeping them locked up. Cats like a large territory and spend most of their time alone. They are really specialised hunters that would normally eat about 10 small animals a day. This takes a lot of time because often the mouse or bird will get away. This means that even though cats love to sleep in the sunshine, they also love to hunt, explore, patrol their territory, climb, look for water as well as food and generally do all the things that millions of years of evolution has drilled into them.

The only way for you to truly answer the indoor versus outdoor question for yourself is for us to look at one of the most important needs of all; the need to express normal behaviour.

Caged birds can't fly or live in a flock, rabbits kept in a hutch can't dig, dogs not taken for walk can't run and chase, and the list goes on. Not being able to behave normally makes our pets miserable, frustrated, bored, angry and ill. So let's look at cat behaviour, how we can make sure they really are happy and how we can make life fun as possible wherever they live.

Gemma Asbury and International Cat Care.

Indoors...

iStock

...or freedom to choose?

Hon Wa Yip c/o International Cat Care.

Chapter 6

THE NEED TO EXPRESS NORMAL BEHAVIOUR.

Naughty? Me? Never!

Shutterstock

Cats are like teenagers. They are difficult to understand, they don't communicate well, they sleep most of the day and they are liable to lash out at you for no apparent reason. But the great news is that if you understand cats and teenagers, you can cope with all of their odd behaviour and maybe even learn to love it!

Now, have you ever had an itch in a place you couldn't reach? Right in between your shoulder blades, for example. You bend your arm right the way up your back but the tip of your thumb stops about a centimetre short of the itchy spot. Ooooohh, it drives you mad, doesn't it? You try and ignore it, but the more you try, the more you keep thinking about it until you can't think about anything else. Eventually your mum finds you crazily scraping yourself up and down the door frame or discovers you have roped her best hairbrush to a wooden spoon and are frantically gouging it up and down your back with a weird look on your face like a cross between sheer panic and total heaven.

This is what it's like for animals who are not allowed to do the things they love or feel the need to do. It's the itch they can never scratch. All animals are born with some behaviours that are what's called 'innate'. This means they are born needing to do something even if they don't know why. Other behaviours are learnt as they grow. For example, an innate behaviour for children appears to be constantly picking your nose, whereas a learnt one is getting a tissue and actually wiping it!

Innate behaviours help animals get a head start because they can do things without needing to be shown. One of the strongest and earliest innate behaviours you see is when animals suckle their mother's milk. Within minutes of being born, calves, lambs, kittens, puppies and human babies all start looking for their first warm drink of milk. They don't think about why and they don't need to but it gives them a great start because they get a full tummy and lots of goodness straight away.

The important thing to remember is that even if we keep an animal in a way that means it doesn't *need* to do something any more, it will still feel the urge to do it. It will still be the itch it can't scratch. For instance you might know that you will never let your cat go hungry, but that won't stop it feeling the urge to hunt.

This is where you'll realise how useful it is to have learnt everything we did about cats in the wild. Pet cats may be more gentle and loving than wild cats but they still have all the same instincts, fears and needs. A few hundred years of being pets doesn't beat out what millions of years of evolution has beaten in!

So let's think about your house, your garden and your cat and all the things your cat will feel the need to do, and then you and your family can decide a) if you can meet all those needs and b) if you can put up with them!

Hunting

There are lots of things that cats do that some people find unpleasant, and one of the biggest ones is hunting. The chances are that if you want a pet, you are an animal lover. So when your lovely pet happily goes and savages other animals to death it can be quite upsetting. Lots of people think that because their cat is well fed and cared for, it won't need to hunt. Do you remember back to the wild cat chapter? We said that even if a cat is eating something it's killed and another animal comes along, it will still try and kill it because it doesn't know when its next chance will be.

The urge to hunt and kill isn't to do with how hungry your cat is, it's just in cats' nature and they can't help it. Not all pet cats do hunt, but you will never know if your cat is going to or not and you can't train it out of it. You just have to accept that if you get a cat, it might kill mice, birds and sometimes even bigger animals like rabbits.

iStock

Some cats are efficient killers...

Cats that do hunt vary in how good they are at it and also what they do with things once they've actually caught them. Some cats are still those perfect stealthy predators like their wild cousins. They stalk in silence, wait for the right moment, judge the distance to perfection and pounce with absolute precision. They bite their prey in the neck to kill it and then they eat every scrap they can (except the really yucky bits!). Now if this is your cat, you may never see this happen or have to deal with anything gross, BUT lots of cats like to involve you and lots of cats don't eat what they catch or, sometimes, even kill it.

Helena Dbalí

...and some just like to play with their prey.

Chapter 6

THE NEED TO EXPRESS NORMAL BEHAVIOUR.

Some cats catch a mouse or a bird and then seem a bit unsure about what they're supposed to do with it. The chances are your cat won't be hungry, so even though it still has the urge to hunt, it doesn't need to eat it. Someone once told me when I got my first cat, Charlie, that if I didn't have mice in the house, I soon would. Sure enough, after a few days Charlie caught a mouse, brought it in alive and kicking, played with it until I caught him, and then wandered off to let the mouse scurry under the kitchen cupboards, never to be seen again! As a cat owner you need to be prepared to rescue random animals from your cat and release them if they are well enough.

Some cats will bring animals in dead and then leave the bodies for you to find. The yuckiest cats eat bits of the animal and then leave heads and bits of intestines and the odd foot for you to clear up. Our cat Brian used to leave the occasional body part in my husband's shoes! So if you want a cat, someone in your family needs to have a strong stomach for dealing with the possible corpses! Owners often wonder why their cats bring these animals home. The simple answer is that it is their home, their territory and where they feel safest to be left in peace with their prize. It's nothing personal!

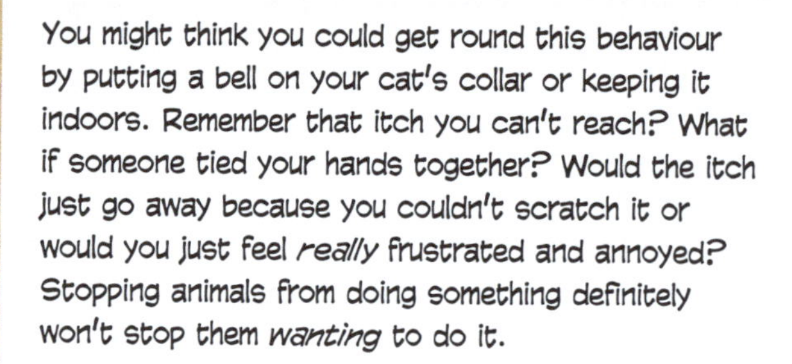

You might think you could get round this behaviour by putting a bell on your cat's collar or keeping it indoors. Remember that itch you can't reach? What if someone tied your hands together? Would the itch just go away because you couldn't scratch it or would you just feel *really* frustrated and annoyed? Stopping animals from doing something definitely won't stop them *wanting* to do it.

It's not OK to deprive an animal of one of its needs because you don't like it or it's inconvenient to you. This is why you have to know about animals before you get them. Lots of cats get given away because people didn't realise they would hunt or scratch the furniture or any of the other things they do. Make sure that's not you!

On the subject of the bell, there are two things to say. 1) Most cats will learn to stalk without it ringing, so bells don't stop them killing animals and 2) for creatures that are as sensitive as cats, having something jangling every time they move is really unpleasant. Please don't put bells on their collars.

You might be able to reduce your cat's need to hunt by giving it toys and playing with it in a way that lets it do the behaviour without hurting anything. Cats love things to chase and some of them even play fetch like a dog. A screwed-up piece of paper makes a great, cheap toy. Wiggling your toes under your duvet or wiggling a finger under a rug or inside a paper bag gives them a great chance to pounce and grab and roll around with some pretend prey. Toys that don't move won't interest your cat. Cats are gripped by sudden movements like their prey make.

All cats love to stalk and pounce, so try to play games that let them do this

Have several toys and change them round them occasionally so they stay interesting. Why not indulge one of your needs – making things – with your cat's need to chase and play and let your imagination go wild!

Scratching

Do your mum and dad like your house? Did they spend a long time arguing in IKEA on a sweaty bank holiday weekend trying to decide on exactly the right settee? When you were little and you were learning to draw and you thought it would be great to draw all over your furniture, did it go down well? So think about how well it will go down if your lovely new cat gouges every corner of your settee, the carpet on the stairs and the beautiful wooden door frames your parents spent hours stripping back through 20 layers of old paint!

Cats scratch things for two reasons. Firstly you might remember we said that they use scent to mark their territory. They have little scent glands in their feet and when they stretch up high to scratch a tree or your table leg, they are leaving behind a little smelly sign that says it belongs to them. Your nose (and every other human's) is way too small and rubbish to smell it, so don't worry if you can't!

Chapter 6

THE NEED TO EXPRESS NORMAL BEHAVIOUR.

Secondly, they scratch things to keep their claws in tip top condition, a bit like filing their nails. Scratching helps to get rid of all the dead, old bits of claw and keep the newest, shiniest and deadliest bits in shape. You can't stop cats from doing it. The best to hope for is to give them things to scratch instead of your furniture.

If you don't want your settee to look like this, you'll need some other things for your cat to scratch.

You can buy scratching posts of all sorts of shapes and sizes. Some have beds and platforms, some have toys attached to them and some of them are huge. You need to know though that you might spend a fortune on a deluxe one and your cat may totally ignore it and scratch your furniture anyway! If you put the post near the place the cat likes to scratch, you can sometimes switch them over. Make sure the post is really sturdy though. Cats need to be able to really push against them without them moving or they won't use them. Get a few different types, like rope or carpet covered.

Some cats like to scratch in different directions, so give them a good choice and you have more chance of saving the house. A lovely piece of driftwood from the beach or a log from the woods might make a perfect scratch post too, so they don't have to cost the earth.

In some countries people have their cats' claws removed so that they can't damage furniture. This is absolutely unacceptable. It stops cats being able to defend themselves, is an unnecessary operation and will leave them very frustrated. If you or your family can't bear the thought of the slightest possible damage, then a cat is not for you.

Hello!

Some cats reach as high up you as they can manage.

Rubbing and Spraying

Rubbing is a cat's way of saying hello. When cats greet other cats they know and like, they rub along each other from their faces all the way along the sides of their bodies. They have little scent glands in their face as well as their feet and by rubbing against each other they get to know their smells and keep their friendship tightly bonded. Your cat will do this to you and your family too. They can't reach your face, so they make do with rubbing round your legs and feet. Some cats rub as high up on you as they can because they secretly want to rub on your face. It's probably best to avoid them rubbing your face though, because they might have just been cleaning their bottoms!

Cats will also rub their faces on things round the house to mark their territory.

Cat greetings like this are really lovely and when they come with a little chirrupy meow as well it can make you feel really happy. It's also very good at making you trip over when you're half asleep in the mornings, so watch out!

Your nose may be too weedy to smell the scents from your cat's feet and face but it is definitely able to smell when cats spray. Spraying is when a cat stands against a wall or a piece of furniture and sprays urine on it. This is its biggest and boldest way to mark its territory. The smell of cat urine is revolting to us but it makes your cat feel safe and at home.

Don't panic, the vast majority of pet cats won't do it, especially if you pick your cat wisely, don't have lots of other cats, you learn about their needs, and have your cat neutered. Spraying is still something you need to know about because when it does happen it is the biggest reason that cats gets abandoned or rehomed. By finding out all the facts and being a good pet owner, you can hopefully avoid it happening in your house. We'll talk more about why they might do it, and also about neutering, in the next chapter.

Chapter 6

THE NEED TO EXPRESS NORMAL BEHAVIOUR.

Life On Their Own Terms

When you talk about cats to other people you will often hear this: 'Urrggh, I hate cats and they ALWAYS come and sit on me.' This is absolutely true and is fascinating to understand. And once you do understand it, you can make yourself much more attractive to cats and a much better cat owner.

Humans are big scary animals and because kids tend to be the loudest, quickest, bounciest, most inquisitive humans, they are the scariest of all to most other animals! People (including kids) who like cats are desperate to fuss them, stroke them, pick them up and cuddle them, so they follow the cat around, trying to grab it, making all manner of odd noises, holding out their hand and staring at them. Cats HATE this. Being followed and clutched is really frightening for them, which is why they usually avoid cat lovers like the plague!

People who don't like cats totally ignore them, never try to touch them and positively avoid making eye contact with them in total fear that the devil creature will come near them. This is like the biggest nicest welcome a cat could have. People who don't like cats are not threatening at all and that is why cats ALWAYS sit on people who don't like them. Brilliant, eh?

If you do get a cat, you need to resist the urge to smother your cat with love. Always be gentle and quiet, don't chase them, follow them round or try to pick them up. When your cat or kitten first comes home it will almost certainly want to hide until it feels safe. Just let it be.

Always let your cat come to you when it is ready. This will get your friendship off to a great start and help make your cat trust and like you as quickly as possible. If you are lucky enough to end up with an affectionate cat, still watch out for when they've had enough of your fuss. If they start flicking their tail or looking over their shoulder at you, stop stroking them immediately and let them go if they want to. If you don't, you could get scratched or bitten and ruin a beautiful friendship. Your cat will be happiest with life on its own terms. Never forget it!

And Finally...

There is a phrase that vets and behaviour specialists use a lot, which is 'environmental enrichment'. This is where we know that having a nice house or cage or hutch to live in is not enough. We need to enrich our pets' environments to make them better and let them behave naturally. You would have already done this a bit with a comfy cat bed and a scratching post, but what else can we do? Yes! Look to the wild.

Cats spend a lot of time finding food and having lots of small meals. They originated from really hot places where they need to explore to find water. When you look at your cat, try and picture the African wild cat lurking in his genes. Make feeding more interesting. Put small meals in all sort of places for him to find (as long as it's in his daily ration!). You can get feeders that make your cat work to get the biscuits and keep him stimulated.

Eleanor and Scott Coultish.

You don't have to spend a fortune on fancy things: a cardboard box is a lot of fun when you're a cat!

Put his water bowls away from his food so that he can eat and then go exploring for water. Most importantly, look at your house and garden from your cat's point of view. Make sure he can climb, hide, jump, snuggle, have space, have peace and quiet and do all the things he feels the need to do. Make sure he is happy and free.

At the end of the last chapter we said that looking at behaviour might help you decide whether it's OK to keep a cat indoors. So what do you think? Is there a right or wrong answer? Personally I think that all cats should have the freedom to go outside. Some cats (a very small number) may *choose* to stay indoors but they should still have the choice. I've seen cats staring longingly out of windows at birds and trees and the beautiful, exciting outside world and I wonder why we think it's OK to deny them. We would never have a dog and make it stay indoors all the time. An indoor life can be very dull and deprive cats of their behavioural needs. You might say, 'Well I live in a high flat' or 'I live somewhere cats aren't allowed out by law, so I would have to keep them in.' Do you remember what I said about needs and convenience? If you live somewhere a cat could never go out, then maybe a cat isn't the right pet for you.

Which cats are scratching that behavioural itch?? This one…

Gemma Asbury c/o International Cat Care.

Karon Stretton c/o International Cat Care.

…or these…

iStock

Shutterstock

Desiree Houtekamer c/o International Cat Care.

Ooh, that got a bit serious again, didn't it? Let's move on from how to keep cats super happy and look at some of the gruesome things we can prevent, to keep them healthy too!

Chapter 7

THE NEED TO BE PROTECTED FROM PAIN, INJURY AND DISEASE.

Every person and every animal gets poorly or hurt from time to time. That's just a simple fact of being alive. I always tell my daughters that their bumped shins and skinned knees are a good sign that they've been having plenty of fun. We already talked about what amazing machines bodies are, and one of their most amazing abilities is how they can heal and recover from injuries and disease. But you'll all know that there are lots of illnesses and injuries that bodies can't cope with and that bodies need extra help with. It's not just about getting help for your pets when they are poorly or hurt: it's very important to find out all the ways you can stop them from getting ill in the first place.

We said before that cats don't usually like being picked up and carried about but it is also important to make sure your cat is used to being handled for health reasons. You will need to check him over regularly to make sure he is healthy and he will need to go to the vet sometimes. If your cat is used to being carefully stroked and examined, he will be less stressed and less likely to struggle and injure you or himself. If you get a kitten, you can get him used to being gently handled from a young age. Whatever age of cat you get, it's also a really good idea to get him used to a cage for travelling.

Lots of cats get very stressed going to the vet or the cattery because they are not used to travelling or being put in a cage. Getting scared cats in and out of travelling cages can be very difficult and with claws and teeth it can occasionally be very dangerous too! If possible have your travel cage around all the time. Leave it open with a comfy bed in it and occasionally put little treats in there to encourage your cat to hop in and out. You can close the lid sometimes for a few minutes so your cat doesn't feel threatened by the cage. Cages that open right up on top are much better for you, your cat and your vet than the ones with the little doors at the front. Trying to squeeze cats in and out of a small opening is no fun for anyone involved! Never take your cat in the car or to the vet not in a cage. Cats can escape really easily if you're just holding them and could get lost or injured.

It's a good idea to go to your vet as soon as possible when you get your cat. He can have a good check over and you can watch and learn how to handle him best. If you are unsure, always get a grown-up to help you.

Let's look at things we can prevent first and then we can have a look at signs to watch out for that might mean your cat is under the weather.

Vaccinations

Depending on your age you may or may not remember having your own vaccinations done. Having them done might involve the slight niggle of a jab but they are one of the biggest life savers of recent times. Once humans started to understand how our bodies fight off diseases, we realised we could help out with some of the most horrible ones.

Your body and most animals' bodies have a super army inside them called the immune system. This army is made up of millions of cells which are always on the lookout for bad bacteria and viruses. They never sleep, even when you do, and they happily lay down their lives to kill the bugs that try to make you poorly. You can think of these cells as your soldiers. They all have different ways of dealing with invaders. Some engulf the germs and eat them and some cells use bullets called antibodies. These soldier cells lurk round your body and when they find an invader they attack it. Now the first time they meet a new invader, they won't have exactly the right sort of bullet to instantly get through the invader's armour, so usually it takes a lot of time and effort from the soldiers to try and overpower the invader and find out what its weakness is. For some nasty diseases the invaders are too strong and our immune cells can't win the battle and don't get the chance to make the right antibodies. These are the diseases which, especially in the past, made us and our pets really ill and even sometimes die.

Vaccinations give your soldiers super, armour-piercing bullets. Humans have studied the viruses and bacteria from many killer diseases. To make a vaccination they get some of the particular bug in a laboratory and they take away its weapons so it can't make you properly ill. The vaccine of weakened bugs is injected into you or given via your nose or mouth. Your soldier cells find the invader without its weapons and they get the chance to examine its armour and work out how to design the right bullets or antibodies. They kill the weakened intruder and they store away the design for the right antibody bullet. This means that when your body gets attacked by the real bug with all its weapons, your immune system is ahead of the game. Your soldier cells immediately make the exact bullet to get through its armour and it's game over before the war even begins. Millions and millions of humans and animals owe their lives to the power of vaccinations.

Different animals get different diseases so they need very different vaccinations. This is why you need to find out all the facts for all the animals you're thinking about owning. There are quite a few vaccines your cat can have, but which ones they need will depend on where you live.

EVERY cat needs to be vaccinated against cat 'flu and a disease with the tricky name of panleukopenia. Lots of people call it enteritis because it's easier to say! And if you live in a country where rabies is a problem, your cat should also have a rabies vaccine. These diseases are very nasty and can make your cat very poorly and even die. Vaccination is essential to keep your cat safe and could also help get rid of these diseases altogether.

International Cat Care.

Flu makes cats' noses and eyes run. Cats with blocked noses quickly stop eating because their sense of smell is closely linked to their appetite. They soon feel miserable and get very ill.

Chapter 7

THE NEED TO BE PROTECTED FROM PAIN, INJURY AND DISEASE.

Talk to your vet about what other vaccinations your cat might need. One of the most common other ones given is for a disease called FeLV or Feline Leukaemia Virus, but there are others your vet might recommend.

Cats and kittens need a course of two or three vaccinations the first time they have them. This makes sure their soldiers make plenty of bullets and are absolutely sure what the invaders' weaknesses are.
After that your cat will need a booster vaccination every one to three years, depending on the vaccine. These must be done to remind the soldiers what to do. These visits to the vet are also excellent because your cat will get a really good health exam, you can get your vet to make sure they are not too fat or thin and you can find out if there is anything new you should know about.

Neutering

Neutering is when your pet has an operation to make sure he or she can't have any babies. This is important to help reduce the number of unwanted or neglected pets. We said before that lots of cats end up without homes and living semi-wild. Although cats usually manage to survive without humans, it is not a pleasant way for them to live because they could be hungry a lot of the time and noone will take them to the vet if they are poorly or get hurt. A lot of people don't have their animals neutered early enough and they end up having accidental babies that could end up without homes.

All cats should be neutered by six months of age but it is common these days to try and do them around the age of four months if possible. This is the easiest way to make sure they never have kittens by accident. It also makes them much less likely to do the unpleasant urine spraying we mentioned in the last chapter.

Being neutered not only stops unwanted babies but it also helps keep animals healthy because it prevents some diseases, like certain infections and some types of cancer. Male cats that are not neutered are also more likely to fight and be aggressive. This makes them get injured a lot but it also makes them more likely to get feline AIDS, or FIV, which can be fatal.

Talk to your vet about when to have your cat neutered and what you need to do to help them get well soon after their trip to the surgery. Once it's done, your cat has a much better chance of being happy and healthy for longer.

Parasites

Parasites are little creatures that live in and on other animals. All animals and humans come into contact with parasites. It's just a fact of life. It doesn't mean you're dirty or bad in some way, it just shows how brilliant parasites are at hitching a ride on the animals they like best. You probably will have heard of the most common parasites: fleas, worms, mites and lice.

All parasites are annoying but some can be very dangerous and some can also pass from animals to humans, so it's very important to make sure you know all about the ones your pet could get. Your vet will be able to tell you about the most important ones where you live and which treatments are safe and effective to use for your pet. Never use products for other animals because you could make your cat ill or even kill it if you use the wrong thing.

Worms and fleas are very common in cats, so let's say a little about them.

Fleas

Close-up of a flea. You can see its huge legs it uses to jump on and off your cat. And your bed. Yuk!

Anyone who says their cat doesn't get fleas is wrong. Even cats kept indoors all the time can get fleas, which come in on clothes and shoes from their human owners. Fleas can't fly but they are some of the best jumpers in the world. They live in the undergrowth and grass and also in your carpets, bedding and furniture at home. They hop onto your cat when they are hungry and bite them to drink their blood. After their meal they hop off again and lay eggs in your house. For every flea you see on your pet there will be a hundred in the house. Yikes!

Fleas are annoying because they cause itching but some cats are allergic to the bites and can be so itchy they tear themselves to pieces and can get infected skin and bald patches. Fleas can also spread diseases and worms, so they can be a massive problem. They can also bite humans.

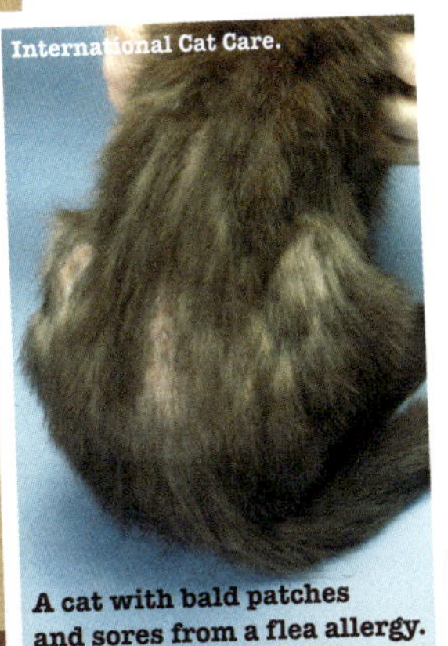

A cat with bald patches and sores from a flea allergy.

Luckily, once again we live in a time where we can get rid of fleas easily and also prevent them from getting a hold in the first place. Use what your vet recommends and keep your treatments up to date and you can ensure your house and cat never get infested!

THE NEED TO BE PROTECTED FROM PAIN, INJURY AND DISEASE.

Roundworms and tapeworms are very common.

Worms

Worms (parasites, not the type you find in your garden!) live in your cat's intestines, so you usually won't know they are there. They lay their eggs there, which come out in your cat's poo and either get swallowed again when your cat grooms itself or get left in the garden to infect other animals. Some worms are just trouble because they take your cat's food from the inside but other worms can cause lots of internal damage as well, so you must treat your cat to kill and prevent worms by following your vet's advice.

Close-up of a roundworm.

These days there are quite a few medicines that treat fleas and worms all at the same time to make life easy. You can even get apps that remind you when to do it!

Bayer Animal Health.

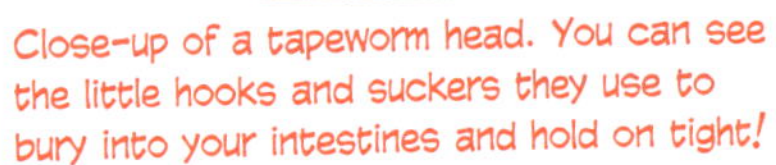
Close-up of a tapeworm head. You can see the little hooks and suckers they use to bury into your intestines and hold on tight!

Microchipping

This isn't a disease, but a good thing to do when you get a cat. Microchips are tiny things about the size of a grain of rice that get implanted under the skin of your pet. Every chip has a different number on it and this gets registered to your details. This way if your pet goes missing it can be scanned with a little microchip reader and you can be reunited really quickly. Lots of people don't like to keep collars on cats, and sometimes collars come off and the tag with your phone number is lost. Having a microchip is the safest and surest way to make sure your pet comes back to you if it gets lost. It's a quick and easy thing to do, so have a chat with your vet about it.

Cats can get used to all sorts if they are well socialised, and some, like my cat Brian, even end up adoring the dogs in their life!

Stress

We mentioned stressed cats before, and stress really can make life miserable for some cats. Because stressed cats do some of the most unappealing things, like spraying or being anti social, they can also make their owners very unhappy too. Cats that are stressed can overgroom and make themselves bald, get gut problems like diarrhoea, which comes and goes, and can also get very serious bladder problems, which can even kill them. Lots of the stuff we've talked about already will have given you pointers about how to keep your cat's life stress free, so let's recap and think of more ways to help.

Brian and Badger, friends for life.

Choosing the right cat or kitten in the first place. The early days and weeks of life for animals is a very important time. It's when animals get 'socialised'. This means they learn about the world they live in and what they are likely to encounter. The earlier that animals meet lots of different things like humans, babies, kids, other animals, hoovers and the general hustle, bustle, smell and sound of human life, the more likely they are to be happy and cope when they are older.

Adopting animals is an excellent thing to do and most adoption centres are experts at matching the right cats to the right households, so definitely look into this because you'll get great advice and also be doing a really good deed.

If you're getting a kitten, here are some pointers:

- NEVER buy a kitten from the Internet.
- Make sure you always see the mother and where the kitten was raised, which should be a normal home environment.
- Ask the breeder about socialisation, worm and flea treatments and vaccinations depending on the kitten's age. If they seem unsure about any of these things, WALK AWAY, no matter how adorably cute the kitten is.
- Get a moggie. The vast majority of cats are still simple moggies of no particular breed. These are most likely to be relaxed pets. The trendy new crosses, which are more wild, are definitely not for you. These cats are not pets and can really struggle to settle in a home environment.
- Look for a kitten that doesn't seem shy or wary of humans and look at what the mum's reaction to you is as well.

Chapter 7

THE NEED TO BE PROTECTED FROM PAIN, INJURY AND DISEASE.

If you can do everything you can to get a laid back kitten or cat in the first place, you'll be way ahead of the game. Here's a recap of the things we've learned so far that will also help reduce stress:

- Only have one cat. If you have two, make sure they've been together from being kittens.
- Always have food, water, litter trays and beds for both cats if you have two and an extra of everything to be on the safe side. Make sure these are spread round the house so your cat can always get away from things he doesn't like but still get to all these important things.
- Let your cat have access to the outside if he or she wants it. Cat flaps are great for this but can sometimes make a cat feel scared if other cats can get in. If this is the case, get a flap that reads microchips. These flaps only let in your cat and noone else's!
- Do not crowd your cat or smother it. Never chase, cuddle, clutch or scare him. Let him come to you every time when he is ready. Don't overfuss him and always watch for signs he's had enough. Don't let visitors annoy your cat either!
- When he first comes home, let him hide. Make sure he has somewhere quiet to go, has access to food, water and a tray and then wait for him to adjust.

Some cat flaps read your cat's microchip so won't let strangers in. This means your cat can relax!

SureFlap Ltd

All these things are pretty easy to do once you understand cat behaviour. By following these simple steps your cat will have the best chance to be happy, healthy and stress free.

Obesity

We said all the way back in Chapter Three how bad obesity is. Fat cats find it very difficult to keep clean and will get horrible, matted fur and dirty bottoms. Being fat, for humans *and* cats, overloads lots of bits of the body like the joints and heart, but cats are also very like humans in another way too. Being too fat is a really common cause of a disease called diabetes in both of us. This disease can be very difficult to treat and can shorten your cat's life, so it's much better to try and prevent it in the first place.

Cats usually get too fat because they are simply fed too much food. But some cats also get too fat because their owners don't understand them. As you learnt already, cats say hello to their owners with rubbing and usually a chirrupy meow. Lots of owners think this is their cat asking for food.

The cat soon figures out that every time it says hello it gets fed, and the vicious circle begins. Sometimes it is just saying hello!

Stick to good rations, use a timer if it helps to avoid the vicious circle we just mentioned, make sure you know what body condition your cat is, and if you are ever unsure just ask. Your vet or vet nurse will always be happy to help you if you think your cat is getting too fat.

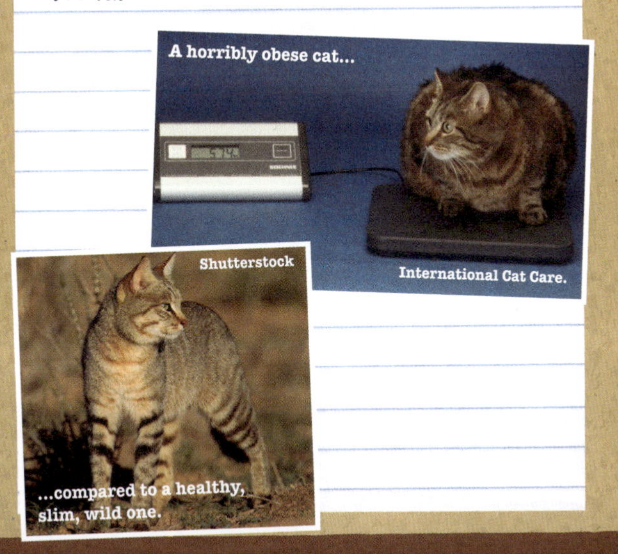
A horribly obese cat...

International Cat Care.

...compared to a healthy, slim, wild one.

Shutterstock

Fur and Funny Shapes!

In general, cats can be pretty low-maintenance pets because they are so independent and healthy by nature. Lots of pets need grooming from time to time to keep their coat and fur healthy and free of mats and tangles. Some pets need way more help than others, but on the whole, good old moggies usually take care of it themselves. As cats get older and their joints get stiff they sometimes need grooming more because they can't reach all the bits they used to be able to.'

Having said this, some breeds of cat have been really changed away from what's natural, both in their fur and their shape. In the wild in nature you will not find very long-haired cats or bald cats, but this is what has happened to some breeds that humans have interfered with.

Cats with very long fur can't keep themselves neat and tidy because the fur is just too tangly. We already said cats don't like being mauled and lots of cats won't tolerate hours of grooming. At the surgery we quite often have to sedate or even fully anaesthetise these cats and completely shave them because their fur is one huge, horrible knot.

Cats are clean creatures and like to keep themselves clean from the tip of their toes...

Shutterstock

...to the tip of their tail.

iStock

THE NEED TO BE PROTECTED FROM PAIN, INJURY AND DISEASE.

At the other end of the spectrum are cats that have been bred to have no fur at all. These poor creatures are prone to sunburn, scratches and lots of skin injuries and often have to be kept inside for their own protection. Long-haired and bald cats are definitely best avoided.

The same is also true for breeds that have become odd shapes. Some cats now have really flat faces like they've run into a wall at high speed. Some have bent ears or really short legs. The trouble is that as soon as humans mess about with nature we cause health problems, and these cats should be avoided.

Just picture the body shape and coat of those healthy happy wild cousins and you won't go far wrong. Moggies all the way!

Is Your Cat Healthy?

The things we've mentioned so far are the big things to be aware of when it comes to having a pet cat, but there are all sorts of other diseases, conditions and injuries that can happen. If you want to learn about every single one then you should do your homework, work hard and become a vet because I haven't got room for all of it here! The best way to keep your pets healthy in general is to be observant and know what's normal so you can spot when things are wrong. Kids are usually way better at noticing things than grown-ups, so put yourself in charge of keeping a watchful eye on the cat!

If you are ever worried take your cat to the vet. Your vet won't think you're silly if there's nothing wrong. It's always better to be safe than sorry.

Things To Watch For In Your Cat

- Bright eyes. No runny eyes or discharge, no swelling round the eyelids or redness. Runny eyes can be because of infections or scratches or even problems elsewhere in the body and should always be checked out.
- Clean nose. No snot or mucus coming from either side.
- Shiny, clean fur. Look out for bald patches, redness and scabs. You don't often see fleas but sometimes if you see lots of little black dots in the fur this could be flea poo.

DETECTIVE TASK.

It's a great detective clue to find. If you are ever unsure whether it is dirt or flea poo there's a clever trick you can do. Comb the specks onto a white piece of kitchen or toilet paper. Add some water. If it's dirt nothing will happen, but if it's flea poo it will turn into a red smudge because flea poo is made of blood. You can amaze and horrify your parents all in one go!

- Comfortable, clean ears. If your cat's ears seem dirty or itchy or you see him keep shaking his head, he could have ear mites. Time for a trip to the vet.
- Body Condition Score or BCS. As we said, it's really good to understand body condition. It's a way of talking about how fat or thin your animals are. The scale is 1-5, where 1 is dangerously thin, 3 is normal and 5 is dangerously obese. Keeping your cat in the right body condition is essential for good health.

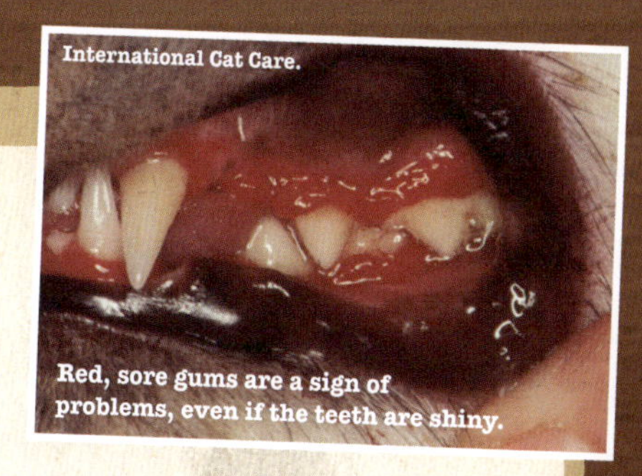

International Cat Care.

Red, sore gums are a sign of problems, even if the teeth are shiny.

- Make sure its nails are not overgrown. Most cats never need their nails trimmed because they look after them so well. Old cats sometimes need more help. If you're ever worried, get your vet to have a look.
- Healthy gums and clean teeth. Your vet will always check your cat's teeth when he has his boosters, but if you notice really stinky breath or if your cat yawns and you see the teeth are brown, get them checked over. Some infections and diseases cause bad teeth or sore gums, so be on the lookout.
- Lumps and bumps. If you find a lump or bump, take your cat to the vet. They can be lots of different things from fight abscesses to tumours. Most are nothing to worry about so don't panic, but get them checked to be safe.
- Look out for odd behaviour. This could be anything from being 'a bit quiet' to hunched, straining to go to the toilet, not moving around as normal, stiff, floppy, sneezing. Basically anything out of the ordinary. If you see any weird or unusual behaviour, get your cat to the vet as soon as you can.
- Normal appetite and thirst. This is the amount your cat normally eats and drinks. Often in cats one of the first signs of problems are changes in appetite or thirst. This could be more *or* less, so it's very important once again to know what's normal. If you do notice any changes in eating or drinking amounts, off to the vet's you go...

THE NEED TO BE PROTECTED FROM PAIN, INJURY AND DISEASE.

Things To Watch For In Your Cat's Surroundings

- Normal faeces (poo) and urine (wee). Lots of cat owners never see their cat's wee or poo because the cat very kindly buries it in the garden. But if you do notice problems like blood in the wee in a litter tray, you need to go to the vet straight away. Diarrhoea is also a definite sign that all is not well. Once again, you'll soon get used to what's normal for your cat. If you don't look, you don't know!

Don't panic about all these things: the more you get to know your cat, the sooner and more easily you'll spot the odd things. The better an owner you are, the healthier your cat will be and hopefully the fewer trips to the vet you'll need. Remember that even though cats are pretty healthy animals usually, you can't just get one and forget about it. Be observant and always, if in doubt, ask your vet: that's why we're there!

Now you have all the facts about wild cats and you know how to keep your pet cat happy as well as healthy. So it must be time for even more fact finding, some virtual reality and some good old maths. Basically it's time to actually answer that crucial question:

Is a cat the right pet for me and my family?

Chapter 8

IS A CAT THE RIGHT PET FOR ME AND MY FAMILY?

In this chapter we're going to crunch some numbers and you're going to have to start investigating facts from some other places besides this book. We're also going to embark on a virtual month of being a cat owner. You might feel silly but it's a brilliant way of checking if you actually do have what it takes to be a dedicated owner of a cat. You can do it all by yourself or you can involve the whole family in the decisions, care and sums.

If you find yourself, towards the end of the virtual month, getting a bit bored with pretending to play with, feed, handle and watch a teddy every day, remember this: a well-cared-for cat can live up to 20 years, so you need to get used to it!

Week One – *How Much??!*

We'll spend the first week finding out some costs to ease you into it gently. Some of these costs are definitely a one-off, like neutering. Things like buying bowls and beds might seem like that too, but bear in mind that over 15-20 years lots of things might wear out and need replacing, so a bit of reserve is always needed. Things like toys are good to rotate or renew so try to get an idea of the cost of say three average toys and allow that amount every month or so.

Now is also the time to consider right from the outset what sort of house or flat you live in and the area it's in. If you were planning to keep your cat indoors and have changed your mind, or if you live in the middle of lots of busy roads, then you might not need to go much further and can avoid the painful maths! So, assuming you think you can provide the right environment for a cat, here's a place to start filling in those numbers. The boxes that are coloured in green are either one-off costs or costs that will only need to be rarely repeated. This will help you and your family get an idea of how much your initial outlay is likely to be compared to ongoing costs. Don't forget though that virtually everything will need to be paid for at the start, so add everything together for your start-up costs!

Things to find out from the pet shop/adoption centre or the internet

Item	Cost £
The cat! Bear in mind that adoption centre costs may include neutering, vaccinating and microchipping. If you're planning to have two cats or kittens remember to multiply this by 2.	
Bowls for food and water. You'll need 1 more set of each than the number of cats you have. (i.e 2 of each for 1 cat, 3 of each for 2 cats)	
Beds. You'll need at least 1 more bed than you've got cats.	
Cat flap. You might need to pay someone to fit this too so check with your mum or dad. Consider a flap that reads your cat's microchip so that no other cats can get in.	
Good-quality food. Talk to a vet before you get your cat. Find out the cost of a bag of dry food or box of pouches and try to estimate how much you'll need each month and the cost. Remember that it's best to offer wet and dry food if you can.	
Timer feeder. This is essential if your cat will be home alone for chunks of time, and allows small, frequent meals as cats prefer.	
Scratching post, 2 if possible, or you can make them, which might be cheaper.	
Toys. Get an idea of how much toys cost, to allow for occasional replacements, and remember that cats aren't interested in things that don't or can't move!	
Litter tray. You'll need at least 1 more tray than cats, even if you're planning to let them go out eventually.	
Total	

Things to find out from your vet

Procedure	Cost £
Microchipping	
Neutering. This will be different for males and females so make sure you find out both.	
Vaccination. Find out what vaccines your cat will need and how often each of the boosters is due.	
Pet insurance. Per month or per year. Ask your vet to recommend a company because some policies can let you down.	
Flea and worm treatment. Cost per month.	
Average consultation cost if your cat gets poorly.	
Total from this and the last page added together	

These numbers may be a bit mind-bogglingly big when you look at them, but that's why I wanted you to do it. There is no such thing as a cheap pet. Obviously you will have worked out that once you're all set up, your monthly costs may be much more manageable, but don't forget about vet's fees and unexpected problems. The brilliant charity the PDSA produce a report every year called the PAW report which looks at how animals are cared for in the UK. They estimate that the average cat costs a massive £17,000 in its lifetime, so you need to be prepared!

Week Two – Handle with Care

This week you are going to spend time and energy devoted to your new (pretend!) cat and getting to know it. It's important to get your cat used to being handled right from a young age if possible. As we said in the last chapter, this will help it feel safe and secure with you and not feel threatened. If it is happy being handled gently, it will be much less likely to scratch or bite you and also less likely to hurt itself thrashing and kicking to get away from you. It will also make it feel less stressed when it needs to go to the vet and be handled by strangers. You'll need to watch for signs it's had enough and let it get away when it wants to.

Remember when you first get a cat that you need to let it come to you. Never chase it or grab it or it will not trust you. Remember what we said about how much cats love people who don't like them? Be like that! Let your cat hide away at first, be around but don't look at it or try to touch it. Offer food and water and wait.

Hopefully if you've picked your cat or kitten wisely or got an adoption centre to do that work for you, then you will have a friendly, sociable and bold cat who will soon be ready to play and come for fuss, but be prepared for the fact it might take time. All animals vary and your cat might not be interested in fuss from you. If your cat is like this, remember to respect its needs and be happy just to watch it and enjoy its company from a distance.

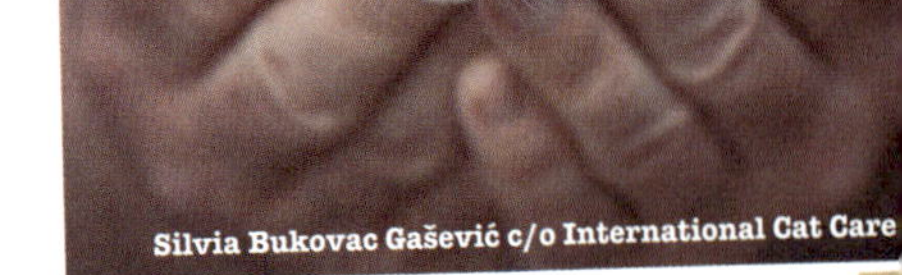
Silvia Bukovac Gašević c/o International Cat Care

Being around a cat is as much about really good observation as it is about handling. Your cat will be in charge of how much physical contact it wants, but this is where your good detective skills will be great. Kids are way more observant than grown-ups. Use this week to think about what you've learnt about cat behaviour. If you do get a cat, you need to learn what is normal for your cat. How it moves, how much it eats and what it prefers. How much it seems to drink most days or how much water disappears from the bowl. What games does it like and what times of day does it seem most sociable? How much time does it spend exploring, sleeping and grooming? The more you notice these things and are observant, the better an owner you'll be because you'll immediately see when something is different.

If and when your cat does come for fuss, and even when you're just watching it, remember to look out for possible health problems. Black dirt, which might be a sign of fleas, bald patches, limping, bad breath, lumps, bumps, clumps of fur or sneezing and runny eyes. If you can get your cat, especially as a young kitten, used to being gently stroked all over, it will help you spot these things. Play games with your cat. Young cats and kittens usually love to play but lots of adult and older cats do too. It can be great bonding time for both of you and fun as well. The more active your cat is, the healthier and slimmer it will be and the less bored and frustrated.

Here is a table to fill in for this week. Once you are in the swing of things you'll find that most of the observation just becomes second nature. It's not really like rabbits and guinea pigs, where you need to go to their cage and handle and check them twice a day. Cats are easy in some ways to have as pets but still keep in mind that you can't just forget about them. You'll probably need to set aside a good hour of your day at least to make sure you do these things right. Oh yes, and don't forget to wash your hands after you've been stroking or handling your cat.

Job/Day	Monday	Tuesday	Wednesday	Thursday	Friday	Saturday	Sunday
Time nearby, stroking and gentle handling if allowed.							
Play time! Be inventive and have fun with your cat.							
Any sign of problems? Time for observation.							

Chapter 8

IS A CAT THE RIGHT PET FOR ME AND MY FAMILY?

Week Three – Fed, Watered and Neat as a Pin!

This week you'll be doing the dirty work! As well as feeding and checking the water is fresh and clean, you'll be learning about cleaning up. It's a rare person who *really* enjoys cleaning anything, so you can be forgiven for not looking forward to this, but it is a huge part of pet keeping. There is no getting away from the fact that some of what goes in has to come out, and it is up to you to clear it up! Handling poo and sometimes wee can make you poorly so you need to make sure you know all about hygiene. Always wear gloves to clean out your cat's litter trays and always wash your hands afterwards.

Of course you still don't have a cat, so this week is about setting aside the time you need as if you had to do these jobs. Why not get your mum or dad to give you a boring or yucky job to do that would take about the same amount of time? Clean the bathroom *including the toilet*, empty all the bins or do the ironing. You'll get an idea of the more boring side of being a pet owner and you'll get massive brownie points at the same time.

Checklist for this week

Job/Day	Monday	Tuesday	Wednesday	Thursday	Friday	Saturday	Sunday
Check/fill timer feeder OR give food in small meals several times a day.							
Clean food bowls or timer daily.							
Clean water bowls and give fresh water AM.							
Check/freshen water and clean bowls if necessary PM.							
Check litter trays 3 times a day, remove poo and wet patches.							
Change all litter and wash tray. This usually needs doing 2-3 times a week but will depend on your cat.		X		X			X
Check toys, and enrichment and replace if necessary (once a week).	X	X	X	X	X	X	

Week Four – EVERYTHING!!

And now for the grand finale. This week you will need to find a couple of hours every day in your hectic schedule to devote to your new pet. Fill in the rather large table below and add in your ongoing costs at the bottom. Most of all, try to enjoy it because if you do get a cat you're going to be doing this for years, not weeks, and probably until you leave home!

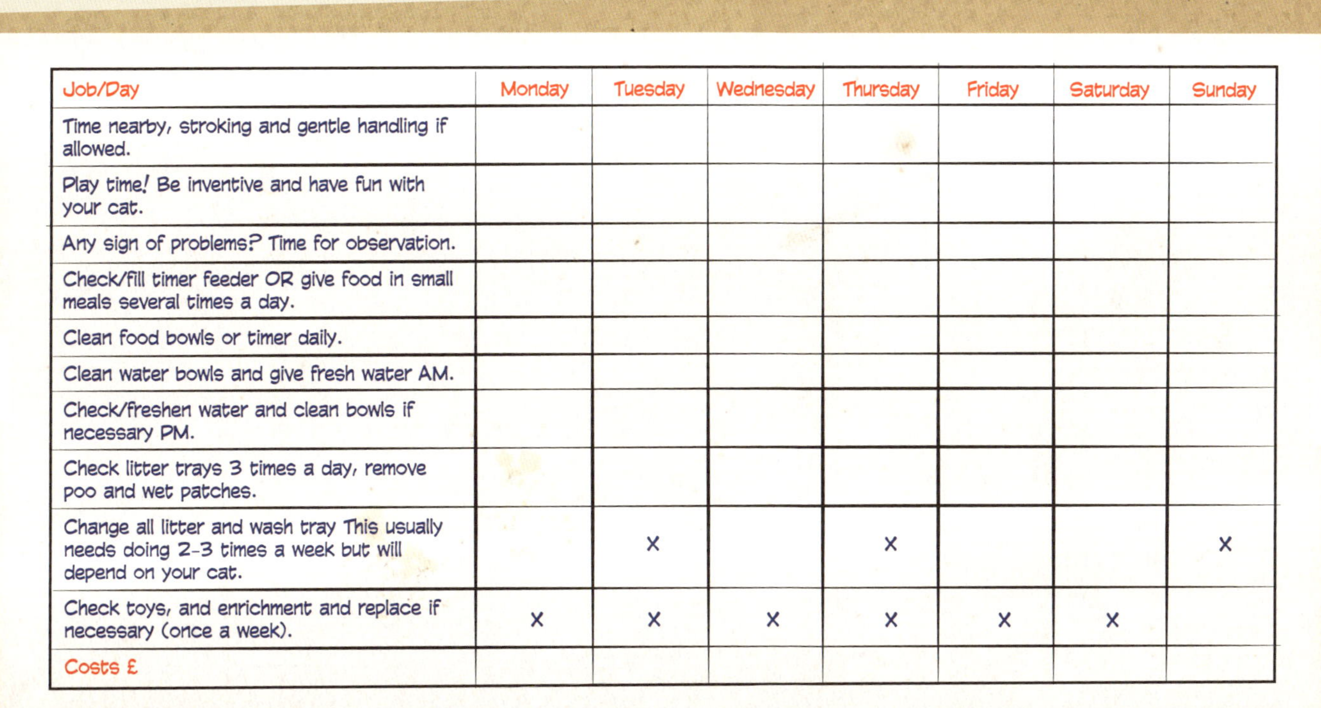

Job/Day	Monday	Tuesday	Wednesday	Thursday	Friday	Saturday	Sunday
Time nearby, stroking and gentle handling if allowed.							
Play time! Be inventive and have fun with your cat.							
Any sign of problems? Time for observation.							
Check/fill timer feeder OR give food in small meals several times a day.							
Clean food bowls or timer daily.							
Clean water bowls and give fresh water AM.							
Check/freshen water and clean bowls if necessary PM.							
Check litter trays 3 times a day, remove poo and wet patches.							
Change all litter and wash tray This usually needs doing 2-3 times a week but will depend on your cat.		X		X			X
Check toys, and enrichment and replace if necessary (once a week).	X	X	X	X	X	X	
Costs £							

IS A CAT THE RIGHT PET FOR ME AND MY FAMILY?

Time for the Family Debate

You've probably been talking to your family about things as you've gone along, but if not, now is the time to do that. You can call a meeting and present your facts, like all the best detectives do. Because now you really do have everything you need to answer that question. **And to answer it honestly.** Things you might want to talk about at your family meeting:

- If you are under 16, someone else in your family will be legally obliged to provide all these things for your cat and they need to agree to that!
- Does a cat, from what you've learned, tick the boxes of what you'd like in a pet? If you thought they were something different, don't be ashamed to change your mind. That's the whole point of finding out all about them: to make the right choices.
- Can your family afford the costs you've found out? Lots of people get embarrassed talking about money, but now is not the time to be shy. If you can't afford it, don't get one.
- Did you have the time, energy and room to provide for all the things your imaginary pet needed? And if so, could you do that for up to 20 years? If you're over the age of about five, you may well be moving on before your cat dies, so your family will need to carry on where you leave off. Are they willing to that?
- Is the whole family on board with the idea?

Hon Wa Yip c/o International Cat Care.

Too many animals get abandoned because people don't find out all the facts.

I hope that after all your hard work you finally get the answer you wanted, but what about if you didn't? Time to ask the next question: what if the answer is no?

Chapter 9

WHAT IF THE ANSWER IS NO?

As we said all the way back in Chapter One, you shouldn't really ask yourself what sort of pet you want, you should ask yourself what sort of pet you can care for properly. The fact-finding you've done up to now will hopefully have helped you work out if a cat is an animal you can keep healthy and, just as importantly, happy. As I said at the end of the last chapter, there is absolutely no shame in finding out the answer is no. That is the point of your mission and the book: to help you and your family make the right and responsible choice. Not only will you have happy pets, but hopefully you'll have pets that make *you* happy too. Very often pets get given away because they were bought on an impulse with no research. In the case of a cat, it can easily happen too. If kept in a crowd or mauled about they can be very unfriendly and unhappy animals and not very nice pets through no fault of their own, simply because they are misunderstood and poorly cared for. The vicious circle starts and soon you've got a scared or frustrated cat weeing all over the house and soon given away.

Karon Stretton c/o International Cat Care.

No? Oh, OK, then.

So if you have done your numbers and learnt your facts and decided that a cat is not the right pet for you or your family, then that is just as worthwhile as deciding to go ahead and buy one. A massive well done either way. You should be very proud of yourself. If you found that a cat didn't tick your boxes or you couldn't tick theirs, it doesn't necessarily mean you can't have a pet; we just need to look at some alternatives, depending on what you were struggling with.

There's a brilliant animal charity we mentioned before called the PDSA and they've come up with a great way to think about having pets, and that is to think PETS! That is Place, Exercise, Time and Spend. Going through these four things for whichever animal you are thinking about is a good way to decide if you can keep it properly. On the PDSA website they have a great tool to help people find the right pet for their own situation, so do have a look at that as well. For now we'll go one step at a time through PETS and see what other pets might suit you best! Remember that this is just a pointer. You will still need to thoroughly research any pet you are thinking of. Just because one animal may need less room or be cheaper to keep, there may be other things about it that might put you or your family off.

Chapter 9

WHAT IF THE ANSWER IS NO?

Place

A cat will share your environment, but as we said, most cats love a big territory to keep them happy. You may have decided your town or village, house or flat is not going to keep a cat happy, so what about the alternatives?

Rabbits and guinea pigs.

These can make great pets, but in lots of ways are harder to keep than a cat. Rabbits and guinea pigs need a lot more space than most people imagine. You can't just buy a hutch and forget about them. They need really big hutches and exercise areas whether they're indoors or out. Having said that, if you didn't want a cat because you're near a busy road, you might find rabbits or guinea pigs a better option because you can build or give them a big, secure, fun place in your garden or house.

Other 'small furries'.

Small furries are things like rats, mice, hamsters and gerbils. Some of these certainly need less space than a cat and would definitely be worth considering. There are some more exotic small furries kept as pets these days, like chinchillas, degus and chipmunks, but some of these need pretty huge cages to let them express all their behavioural needs, so be very careful to do your research before you decide.

Dogs.

These may be very different to what you were originally considering and in some ways need lots more space than a cat but are still worth thinking about. Like a cat, dogs live in your place but they go out with you to exercise. This means that dogs can be happy without the constant freedom and large territory that cats like. Remember though that dogs have other very complex needs, so make sure you can provide for those too.

Fish.

Fish are very calming, beautiful animals to watch and are very popular with lots of people. A fish tank doesn't need to take up much room, but if you look into fish always consider how much room they would like, because a tiny bowl can be just as bad as a trapped cat. Please, if you do look into keeping fish, find out about where they come from. Some will be taken from the wild and this could be very damaging to the place they come from and the animals and plants that live there.

Exercise

In the case of a cat, exercise is kind of included in space because you need to give them the space and enrichment to exercise when they want to. I like to think of this 'E' as energy instead. It takes a lot of energy to care for pets properly. It could be the energy you need to think about all the jobs you need to do on top of your already busy social schedule. If, when you did your virtual month, that all seemed like a lot of hassle, what else could you consider?

All pets need some commitment from you, but some need less interaction or different interaction, which might be easier. Rabbits, guinea pigs and small furries may be a good option and will exercise themselves given the right environment, but that will take time and energy on your part to get right in the first place. Fish may be a good option for the same reasons. Dogs in general need lots of energy devoted to them and their exercise, so if you're a bit on the lazy side, I would definitely steer clear of them!

Time

In your virtual month, especially in the last week, you should have found that your cat needs a good 1–2 hours of your time *every* day and sometimes more. This might not sound a lot to many people, but when you actually have to do it day in and day out it can quickly become difficult to find the time. You probably won't be surprised to hear that lots of non-pet-detective owners don't realise the time needed until it's too late, and pretty soon we're back to that neglected and forgotten cat or an abandoned pet.

Time is very precious to lots of people. We live in busy times. Lots of parents work, lots of kids do a million after-school activities and weekends disappear in the blink of an eye. Trying to find a spare two hours every day can be a massive headache for any family, even if you share the work. All pets need some time commitment from you and it's definitely worth considering right now if you found time an issue at all in your virtual month. It's better not to have a pet than to have an unhappy, badly looked-after one.

Time and energy go hand in hand. If you struggled to find the time for a cat, then many pets like dogs, rabbits and guinea pigs will be the same. As we said with the small furries, in some ways they may need less time. It will vary depending on the animal, and as always research will be important. All the small furries will need some time for cleaning and feeding and again that will vary depending on the animal and the size of their cage, how they live and how dirty they are! Fish could well be another good alternative, but as with the energy required, some like marine fish might need more time than others to keep their tank exactly right.

Time is a precious commodity, and so is money...

Spend

The cost of keeping pets is probably the most massively underestimated thing of all when it comes to owners. Stuff like bedding, toys and cages are easy things to work out and think about but people always forget things like vet's bills, vaccinations and, quite shockingly, the cost of food. Feeding an animal for 2–20 years can make a big hole in your wallet!

The PDSA PAW report in 2015 asked lots and lots of owners how much they thought it would cost to look after a dog, cat or rabbit for the whole of its life. Most people thought it would cost £1000–£5000 for a dog, when in fact, depending on the breed, it can be up to a staggering £31,000! Most people gave the same answer of £1000–£5000 for a cat, but as we said before, the average is actually often a whopping £17,000. The average rabbit costs £9000, but when owners were asked what they thought it would be, nearly all of them said less than £1000.

You can see why people are shocked when they actually get the animals! Money is another big reason pets get given away. As I said, some people are shy when it comes to talking about money, but if you're thinking of getting a pet, it's absolutely vital you find out how much it is likely to cost and make sure your family can afford it.

From these numbers you'll have already guessed that if a cat was too much of a strain moneywise, then dogs are certainly out of the question.

Rabbits, guinea pigs and small furries.
In general, the smaller an animal is, the shorter its lifespan. For instance, rabbits can live up to ten years, guinea pigs around five years, hamsters two years and mice up to a year. The smaller the animal, in general the smaller the cage you need, so on average the smaller ones should be cheaper than a cat to keep. Some of the more exotic ones will be very different though, and many of these animals shouldn't be kept alone so do your research.

Fish.
Fish will be really variable depending on how many and what type you have. They have hugely different needs for things like water temperature, food and care. A couple of coldwater, freshwater fish will definitely be cheaper than a cat, but a state-of-the-art tropical underwater heaven may not!

What if You're Worried You Can't Manage Any Pet?

Above all, please believe me when I say that it is better not to get a pet at all than to neglect one. Being a responsible pet owner is all about making the right choice and sometimes that means not getting one at all. Try not to be too downhearted. Look at all your options and also think about ways things might change with time. When I was very little we had no spare money at all, but as we got older my mum and dad worked hard and trained and got different jobs and as time went by we found we could get a dog. Trust me, I pestered for a *long* time before that actually happened!

If you have friends who have pets, ask if you can spend time with their animals and help them with the jobs. You never know, they might be a bit bored with it and might love to have a helper. Just being around animals is a brilliant feeling, so you might need to just take small opportunities when they come along.

Talk to your mum and dad about fostering pets. There are so many unwanted pets that lots of adoption centres often need people to foster animals while they are waiting for a home. This might mean having rabbits, small furries and even dogs and cats if you can manage, but just for a short time and often with help from the charity with the costs. You'll get some animal time but also be doing a really good deed too.

You could also find out about charities like Hearing Dogs for Deaf People. Some of the centres ask people to care for a dog in the evenings and over the weekends while it's being trained. This means if your mum and dad are at work during the day, the dog just comes to you for the times you're all at home. They have lots of different volunteering options on their website, so you might find something that is perfect. A temporary arrangement might be a great compromise for you and your family.

So here we are, almost at the end of your journey into the wonderful world of cats and being an A-grade owner. All that's left to say is...

Chapter 10

WELL DONE!!!!

Shutterstock

WELL DONE, DETECTIVES!

By the time you get to here you will have worked hard and learned exactly what it takes to be an excellent pet owner. Let's think about all the amazing things you have done and achieved since page one:

- You've learnt about how animals can make us happy and what it means to be a responsible pet owner, including the serious stuff about the law!
- You've found out how wild cats live, what makes them happy and what keeps them healthy.
- You've found out what food is best for a cat, how much food they need and how to tell if they are too fat or too thin.
- You've discovered just how much room a cat needs to live happily and what sort of environment will keep them safe.
- You've learnt that cats don't usually need or even like to have lots of other cats around. This can make them feel overcrowded, threatened and unhappy.
- You've found out that it's very important to let animals behave normally. A cat will need to be able to play, run and jump, scratch, groom, hide, mark its territory (and its owners!), explore and maybe hunt and kill other animals.
- You've learnt about obesity, cat 'flu, fleas and worms, neutering, microchipping and all the things you need to look out for to spot a poorly cat before it gets too bad.
- You've done extra research, spoken to vets and nurses, been to pet shops, looked online, done lots of maths and maybe even made your whole family sit down together to discuss this whole pet-owning business.
- Hopefully you've even gone the whole hog and made yourself feel a bit silly wandering round the garden or the house pretending to do stuff to a pretend cat.

THAT is a very impressive list of things and that is why you should feel very proud of yourselves. I am ecstatic that you bought or borrowed this book and read it and I am very proud of you. You may feel like kids are ignored sometimes and you might feel sometimes like noone really listens to your opinions, but I'm going to let you into a secret: you kids can change the world. Let's face it, grown-ups have messed up pet keeping for hundreds of years. They think they're too busy to do research and lots of them think they know everything already!

Just imagine if you told your mum or dad all the things you'd learned about how to actually care for a cat. I bet they would be astounded. I bet you could teach *them* some things. They might think it's fine to keep them indoors in a big crowd of cats. But now YOU know better. You can tell your cat-hating aunty why cats simply love her so much. You might not feel like you can change the world, but if all the children in the world learnt what you have and really took it on board, then pet keeping would transform overnight. Old, wrinkly vets like me would smile and put our feet up because our workload would halve in an instant thanks to all those well-cared-for pets.

I've told you repeatedly that you should never ask yourself what sort of pet you want, but now we're at the end of your journey I think it's time for you to do just that. You see, it's very important that you ask yourself what sort of animal you can care for properly, but you do also need to consider what you're looking for in a pet, because you and your family need to be happy too. You might have discovered that you can perfectly care for a cat, but if you wanted a pet that would come for a cuddle when you wanted it to, then a cat might never make *you* happy! Owning pets is a team game, so make sure you all talk about your options.

If a cat isn't right for you, have a look at the other books in the series and keep learning. All animals are fascinating, even if you don't end up with one.

If you've decided a cat would make you happy and that you can keep one happy and healthy, hang on to the book. Unless you've got a brain the size of a planet, you might want to remind yourself of things when you get your new bundles of joy. When you're deciding where to get your cat, please remember adoption centres and the importance of rehoming animals. There are, sadly, always plenty of wonderful animals looking for a loving home.

If a cat now seems like the worst choice in the world, then pass the book on. You could give it to a friend or sell it for some pocket money. If you've got a friend with a cat you don't think is being very well looked after, you could slip the book into their school bag as a nudge in the right direction. If that friend makes a change to improve their cat's life, then you've taken one more step towards changing the world.

At the start of the book we said that living with animals can be wonderful. I hope the books in this series will help to guide you and your family to what will be a fantastic friendship and a time that you will look back on with big smiles and a mountain of happy memories.

All that remains for me to do is to tell you again how brilliant you are and to award you with your detective certificate. Proof that you now know pretty much everything there is to know about the needs of a cat.

WELL DONE!

Congratulations. Now if you'll excuse me I need a quick catnap.

Adrian Tan c/o International Cat Care.

THIS IS TO CERTIFY THAT

Sophie Richmond

HAS LEARNT PRETTY MUCH **EVERYTHING**
THERE IS TO KNOW ABOUT CARING FOR A CAT
AND BEING AN EXCELLENT PET OWNER!